T0140101

# Studies in Computational Intelligence

## Volume 733

**Series editor**

Janusz Kacprzyk, Polish Academy of Sciences, Warsaw, Poland
e-mail: kacprzyk@ibspan.waw.pl

*About this Series*

The series "Studies in Computational Intelligence" (SCI) publishes new developments and advances in the various areas of computational intelligence—quickly and with a high quality. The intent is to cover the theory, applications, and design methods of computational intelligence, as embedded in the fields of engineering, computer science, physics and life sciences, as well as the methodologies behind them. The series contains monographs, lecture notes and edited volumes in computational intelligence spanning the areas of neural networks, connectionist systems, genetic algorithms, evolutionary computation, artificial intelligence, cellular automata, self-organizing systems, soft computing, fuzzy systems, and hybrid intelligent systems. Of particular value to both the contributors and the readership are the short publication timeframe and the worldwide distribution, which enable both wide and rapid dissemination of research output.

More information about this series at http://www.springer.com/series/7092

Piotr Kosiuczenko · Lech Madeyski
Editors

# Towards a Synergistic Combination of Research and Practice in Software Engineering

*Editors*
Piotr Kosiuczenko
Faculty of Cybernetics, Institute
  of Computer and Information Systems
Military University of Technology
Warsaw
Poland

Lech Madeyski
Wroclaw University of Science
  and Technology
Wrocław
Poland

ISSN 1860-949X                    ISSN 1860-9503  (electronic)
Studies in Computational Intelligence
ISBN 978-3-319-87974-1         ISBN 978-3-319-65208-5  (eBook)
DOI 10.1007/978-3-319-65208-5

# Preface

The term "software engineering" was coined at a NATO Software Conference almost 50 years ago. Since then a visible progress has been made in both research and practice in software engineering.

This book is devoted to a synergistic combination of research and practice in software engineering and contains 15 selected contributions. Actually, it is already 19th in the series of books on software engineering prepared under the auspices of Polish Information Processing Society (PIPS). Software Engineering Section of the Committee on Informatics of the Polish Academy of Sciences decided to support these efforts as well. The books are devoted to various topics in software engineering and are addressed to researchers as well as practitioners, engineers, managerial staff from the IT companies and government. To disseminate the results contained in this series, the authors of the chapters (among other researchers and practitioners) present their contributions at KKIO. It provides a forum for presentation of research results, scientific challenges faced by the industry and scientific methods that could address them. It is also a platform to initiate cooperation among researchers and between academia and industry.

The fact that this time it was already 19th edition of the book under the auspices of PIPS shows continuing interest in software engineering.

This year the spectrum of topics was wider and covered also topics concerning real-time systems engineering and education in software engineering. We selected 15 of 43 chapters based on relevance and the value of scientific contribution. This brand-new book, including the selected chapters, was published by Springer in the well-established "Studies in Computational Intelligence" series.

Selected chapters concern:

- languages and tools for software development,
- software development processes,
- modelling and verification,
- education in software engineering.

In the first category, there are seven chapters on topics such as: costs of computing unit redundancy; a domain-specific language for interactive programming exercises; managing software complexity by similarity patterns; tools for validation of class diagrams and ensuring exception safety; testing of time-dependent, asynchronous code; an automatic processing of dynamic business rules.

In the second category, there are four chapters on: continuous test-driven development and its empirical evaluation in industrial settings; enterprise architecture modifiability analysis; the influence of business analysis techniques on software quality characteristics; female leadership in IT projects.

In the third category, there are three chapters on: modelling and verification of real-time systems; access control model for mobile systems; modelling and simulation of computer networks.

In the last category, there is one chapter on a scrum-based framework for organizing software engineering courses.

There are people who helped in the preparation, publication and dissemination of this book. We would like to thank: authors of the contributions, the referees for helping us in the selection process, and PIPS for continuous support for this series. We would like to express also our gratitude to prof. Janusz Kacprzyk, the editor of the "Studies in Computational Intelligence" series, and Dr. Thomas Ditzinger from Springer for their interest and support.

We sincerely hope that this book will be a valuable reference work in software engineering research and practice.

Warsaw, Poland                                                              Piotr Kosiuczenko
Wrocław, Poland                                                             Lech Madeyski
June 2017

# Contents

# Temporal Costs of Computing Unit Redundancy in Steady and Transient State

Jacek Stój and Andrzej Kwiecień

**Abstract** Industrial real-time computer systems are designed to operate with high reliability level. It is most often achieved by the use of redundancy. Additional elements introduced to the system are supposed to make it invulnerable to failures of their redundant components. Implementation of redundancy entails significant extra costs. The financial costs are quite obvious and easy to estimate they are associated with the need to deliver, configure, program, service and maintain additional devices and communication interfaces. There are however another costs which are not so evident because they derive from the system operation and data processing. Those are the temporal costs of redundancy that influence the real-time systems characteristics and in the end may determine the system usability.

**Keywords** Networked control systems · Real-time constraint · Real-time system · Redundancy · Reliability · Steady state · Synchronization · Temporal cost · Transient state

## 1 Introduction

Computer systems implemented in industry are supposed to deliver means of supervision and control of industrial processes the object of control. The object is equipped with appropriate sensors to deliver to the computer system information about the object state. The computer system analyzes the current state and react to it according to the implemented requirements. It also influences the object by means of actuators. The systems are usually distributed and therefore called Networked Control Systems NCS [1, 2] In NCS systems it is defined how much time is available for the system

J. Stój (✉) · A. Kwiecień
Faculty of Automatic Control, Electronics and Computer Science,
Institute of Informatics, Silesian University of Technology, Gliwice, Poland
e-mail: jacek.stoj@polsl.pl

A. Kwiecień
e-mail: andrzej.kwiecien@polsl.pl

© Springer International Publishing AG 2018
P. Kosiuczenko and L. Madeyski (eds.), *Towards a Synergistic Combination of Research and Practice in Software Engineering*, Studies in Computational Intelligence 733, DOI 10.1007/978-3-319-65208-5_1

response, i.e. to react to the changes in the object state. To secure the satisfaction of those temporal requirements, the system must consist of only real-time components, i.e. components which operation is defined and possible to describe in the time domain [3].

Apart from the temporal parameters, the reliability of industrial systems is of great importance. To achieve high reliability level most often redundancy is used. It means that some of the system elements are multiplied (most often duplicated or triplicated) to immunize the system to failures of corresponding redundant components. Application of redundancy changes not only the reliability of the system, but has also a significant influence on the temporal characteristics of the system. Every element in the system has to be serviced which consumes time. In other words, introduction of redundancy to a given system, expands the response time of the system and makes it to operate with greater delay [4].

While the expansion of the response time caused by redundant elements is taken into consideration during new systems design, it is not always so when systems are developed. Introduction of redundancy to an existing system should enforce an analysis of the system operation in the time domain. Making even seemingly unimportant changes in the system, may effect in significant response time expansion and as a result real-time constraints violation. If that is the case, the system with redundancy will be more reliable but inapplicable. If it was exploited nevertheless, then a danger of missing a deadline may occur, which may lead to serious damage to the environment, the industrial plant or cause harm to human beings.

The temporal characteristics of NCS with no redundant components (Non-Redundant System NRS) usually do not change in time. In fault-free conditions the maximum time needed to perform a given task is constant (worst case is considered here). When a fault occurs, the NRS system operation depends on the severity of the fault. If it is considered to be small, i.e. the faulty element is not a critical one, the system continues to operate with possibly decreased functional characteristics like efficiency or with some functionality unavailable. On the other hand, on critical element fault, the system starts the shutdown routines switching the object to the safety position.

In systems with redundancy (Redundant System RS), the principle is quite different. Redundancy is applied to the most critical elements of the system. When one of those fails, the system continues its operation basing on the corresponding redundant element. It means that the hardware configuration of the system may change in time as a result of fault occurrence. The change has to be serviced in the software in a timely manner.

From the point of view of temporal costs of redundancy, two states of the RS systems should be distinguished: *steady* and *transient* state. The steady state would refer to fault-free condition of the RS system. Whereas the transient state would be during the switchover between corresponding redundant elements of the system. Taking a system with redundant communication bus as an example, the steady state of the system would refer to conditions when the communication is executed on the active bus with no disturbances caused by fault occurrence (assuming that only one out of two available buses may be active at one time). The transient state starts at the moment of detection of a bus fault such as communication bus damage. Then the systems switches over to the redundant bus and sets it active. It may require performing some additional routines that starts up the communication process. With the end of those routines, the system state changes back from the transient to the steady state (with no redundant bus anymore as the previ-ously active bus is down). As previously mentioned, when building or developing real-time systems with redundancy, the redundancy temporal costs may influence the system usability from the point of view of satisfaction of the real-time constraints. Therefore, it is crucial to know the source and size of the additional delays introduced to the computer system together with redundancy. In this paper one of most common redundancy architectures is analyzed architecture with redundant computing unit.

## 2 Related Work

The contemporary issues related to networked control systems NCSs are in-cluded in [5–10]. Various real-time communication mechanisms are present in industrial applications: token networks [11], Producer-Distributor-Consumer [12], Master-Slave [13]. There are also another protocols that cannot be strictly assigned to any the above groups like EtherCAT [14, 15] or Profinet [16, 17] both based on the Ethernet standard. The security in industrial networks is considered in [18] also using new (in industrial applications) techniques of virtualization [19]. Various considerations on their temporal characteristics of computer systems are included in [20–26].

There are many works on redundancy (e.g. [27–31]) and the triple modular redundancy too (see: [32]). The fault-tolerance is discussed in many papers, e.g. [33–35]. Moreover, a few interesting thoughts about using multiplied communication interfaces are presented in [36].

The IEC working group produced also the IEC 62439 standard that defines redundancy methods applicable to most industrial networks, and which differ on the topology and the recovery time. The standard describes hot-standby switchover redundancy like the Media Redundancy Protocol MRP (IEC 62439-2, [30, 37]), the High availability Seamless Redundancy HSR and Parallel Redundancy Protocol PRP (IEC 62439-3, [38, 39]). The last to are active redundancy approaches that work without reconfiguration timeouts when a single failure in one of its two redundant network structures occurs [40]. The above refers however to Ethernet-based NCSs.

## 3 Temporal Costs of Redundancy

The temporal costs of redundancy are different for the steady and the transient state. For the steady state the time needed for the service of the redundant elements should be taken into account. While in the transient state, time consumed by switchover routines is most significant.

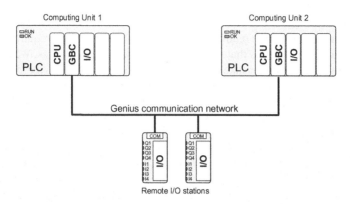

**Fig. 1** The considered system with computer unit redundancy

In this paper a system with redundant computing unit CU is considered (see: Fig. 1). For communication Genius network is used a token passing network with *implicit token, genius.*

The CUs are based on PLC controller (GE Fanuc Rx7i PacSystems including IC698CRE020 CPU module) equipped with Genius communication network controller IC687BEM731. Through the communication network two remote I/O stations were connected. Both of them were based on GE Fanuc VersaMax Genius interface module IC200GBI001 and IC200MDL841 I/O module with 75 bytes of data. The Genius network operated with 153.6 kb/s std.

The considered system with redundancy is shown in Fig. 1. The basic system (without redundancy) would be a system without one of the computing units. The cost of redundancy implementation will be defined by comparing the response time of the system with and without redundancy for both steady and transient state. For that purpose the cost of redundancy index is defined as quotient of the response time in system with redundancy divided by the response time in system without redundancy:

$$C_R = \frac{T_{ResR}}{T_{Res}} = \frac{T_{Res} + T_{CR}}{T_{Res}} \,, \tag{1}$$

where:

$T_{Res}$—the response time in system without redundancy,
$T_{ResR}$—the response time in system with redundancy,
$T_{CR}$—temporal costs of redundancy.

In the following subsection some theoretical analysis of the temporal characteristics of systems with redundancy is summarized.

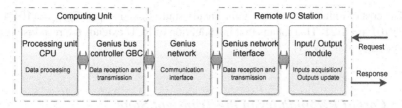

**Fig. 2** Data flow associated with the response time in NCS

## 3.1 Basic Configuration

The considered system without redundancy consists of one computing unit and two remote I/O stations. The response time in the system is the sum of the following: acquisition of the input signal $T_{In}$, transmission of the input state to the computing unit, input processing in the computer unit, transmission of the response to the output station, output update $T_{Out}$ (see: Fig. 2).

According to the manufacturer documentation, in Genius communication network at least 3 communication bus cycles are required for the transmissions of inputs and outputs for the response generation. Additionally, at least two computing unit cycles are needed for input data processing [42].

The total time of the response time for the analyzed system is $T_{Res} = 101.4$ ms for the computing unit cycle equal 3.0 ms and Genius communication bus cycle equal 16.4 ms. The details of the analysis and experimental research that proves it may be found in [42] and are out of the scope of this paper.

## 3.2 Cost of Redundancy in the Steady State

The computing unit redundancy, as shown in Fig. 1, introduces to the system two kinds of delays. First of all, the redundant computing unit RCU is connected to the communication network and needs some time for data exchange. Therefore, the communication bus cycle extends by the transmission of the input and output states. Moreover, there should be some communication with the basic computing unit BCU to enable diagnostics of the operation of both computing units. That extends both the communication bus cycle and the computing unit cycles. Systems with additional synchronization modules installed in the computing units are not considered here. Nevertheless, they also extend the communication unit cycles.

For the analyzed system, introduction of additional computing unit to the network increases the communication bus cycle from 16.4 to 23.5 ms. Assuming that the computing unit cycle extension is relatively small (and may be considered as negligible) the cost of redundancy in steady state is equal $T_{CRCU} = 21.3$ ms (as mentioned above three communication bus cycles are required for response generation).

The cost of redundancy index in the steady state $C_{RCUs}$ is equal $C_{RCUs} = (101.4 + 21.3)/101.4 = 1.21$. That means that introduction of RCU extends the response time in the steady state by 21.

### 3.3   Cost of Redundancy in the Transient State

The transient state in the system with computer unit redundancy takes place when a fault of currently active unit occurs. The state lasts until the redundant computing unit RCU takes over the industrial object control from the faulty computing unit BCU (hot standby architecture is considered). The response time of the system extends by the time of BCU fault detection and time needed for realization of the control *take-over* routines.

In the Genius network, a subscriber is considered to be unavailable on the communication bus when it does not transmit any data for 3 network cycles. In other words, 3 cycles must be realized before a fault of one of the computing units is recognized.

In case of the considered configuration with the network cycle equal 23.5, the fault is detected after 70.5 ms. After that time, the system switches to the redundant computing unit. In this case, the response time in transient state is the response time of the system with redundancy in the steady state plus the time of fault detection: $101.4 + 21.3 + 70.5 = 193.2$ ms. The temporal cost of redundancy index in the transient state is as follows:

$$C_{RCUt} = \frac{T_{ResRCUt}}{T_{Res}} = \frac{T_{ResRCUs} + T_{CRCU_{FD}} + T_{CRCU_{SW}}}{T_{Res}} , \qquad (2)$$

where:

$T_{ResRCUt}$—the response time in the transient state,
$T_{ResRCUs}$—the response time in the steady state,
$T_{CRCU_{FD}}$—time needed for the fault detection,
$T_{CRCU_{SW}}$—duration of the switchover routines.

The $C_{RCUt}$ in the considered system is $193.2/101.4 = 1.90$. That means that during after fault of the active computing unit the response time is almost two times as big as in the system without redundancy.

The control *take-over* routine is negligible in this case. In case of the Genius communication network, the remote I/O stations determine which of the computing units controls the object. When they detect a communication fault with the active unit, they switch to the redundant unit with no delay. It is possible because the I/O stations collect output data from both units during the steady state. Therefore the output vector of the redundant unit is known at the moment the fault detection of the active unit.

## 4    Test Bed

The calculated temporal costs of redundancy presented in the previous section were verified during some experimental research. For that purpose the laboratory NCS with hardware configuration as described above was used.

### 4.1    The Response Time of the Basic System

The response time is the time needed for input acquisition realized by the remote I/O station, input data transfer from that station to the computing unit (PLC controller), response generation (execution of some code in the computing unit), output data transfer (which includes the generated response) to the remote output station and finally physical output update (see: Fig. 2). To measure the response time, some additional device would be necessary. It would generate the input signal, detect the systems response and calculate the delay.

The authors however, used another method that changed the scenario but let them measure exactly the same time. During the experiment one of the system outputs was electrically connected with one input (with a jumper). Therefore, any change of the output was visible by the change of the input. The computing unit was setting periodically the output to a given state. Then it measured the time until the input changed to the same state as the output. The input change occurred with a delay associated with data transfer to the remote station and back again as shown in Fig. 3. The measurement included the following sequence: sending the output data to the remote I/O station, output update realized by the remote I/O station, setting of the input to the output state (done by the electric circuit), input acquisition, input data transfer to the computing unit and detection of the input change.

The response time was measured 10,000 times. On the histogram in Fig. 4 all recorded samples are included. The y-axis is the number of samples grouped by the response time value divided into 5 ms groups.

The maximum response time recorded during the experiment was 91.2 ms. It is less than the calculated maximum response time what corresponds to the theoretical analysis.

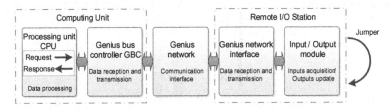

**Fig. 3**  Measurement of the response time during the experimental research

**Fig. 4** Response time
measurement results in
system without redundancy

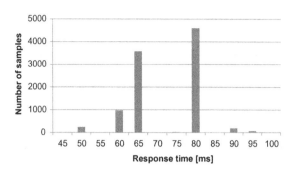

In the above histogram, 4 sets of samples may be distinguished. They belong to the groups of 50 ms response time, 60–65 ms, 80 ms and 90–95. It is worth noting, that the bus cycle duration measured during the experiment was around 14 ms (the maximal calculated bus cycle time is 16.4 ms). The indicated sets fit into the multiplication of that time.

## 4.2 The Response Time in the Steady State

The response time in the system with computer unit redundancy was measured using the same method as described above. The results are shown in Fig. 5. The maximal response time registered during the measurement was 107.8 ms.

It is worth noticing again that the samples fall into three sets which are placed around 70 ms, 90 ms, and 110 response time (only few samples in the last case). It corresponds to the measured bus cycle which was around 20–21ms (obtaining the bus duration with greater precision was not possible).

The redundancy costs according to the measurement is as follows:

$$C_{RCUs} = \frac{max(T_{ResRCUs})}{T_{Res}} = \frac{107.8}{91.2} = 1.18 , \qquad (3)$$

**Fig. 5** Response time
measurement result in the
steady state

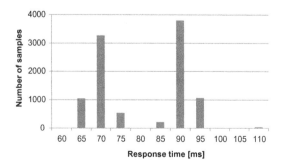

It means that the response time of the considered system after application of computer unit redundancy will have 18% times greater response time than the system without redundancy.

## 4.3   The Response Time in the Transient State

The transient state in NCS with redundancy is associated with the system behavior after a fault occurs. Therefore, for the needs of the experiment, fault simulation is required. One of the solutions is to cut off the active computing unit from the communication bus making it unable to react to requests. After that happens, the remote I/O stations detects the fault on the network and switches to the redundant computer unit which starts to control the object. Hot standby redundancy is considered here.

The above solution may seem to be inappropriate as the simulation of computing unit fault is done by influencing the communication bus. Nevertheless, disconnecting the computing unit from the communication bus have the same result as unplugging the power supply from the computing unit or stopping the communication module to operate the network subscriber is not present on the network and it stops receiving and transmitting data.

The fault simulation requires to change the response time measurement method. It can no longer be done internally by the computing units themselves. Additional device is needed. Its task is to generate a request (some change of the system inputs state) and measure the time in which the system reacts to that request. The test bed is shown in Fig. 6. Additional devices needed for the fault simulation (micro PLC) and response time measurement (PXI measurement computer by National Instruments) are distinguished there.

Faults may occur at any time. Therefore, in the experimental research they also should be simulated in various moments in relation to the generated request. It was

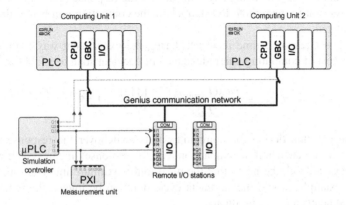

**Fig. 6**  System diagram for transient state experimental research

**Fig. 7** The response time in
transient state

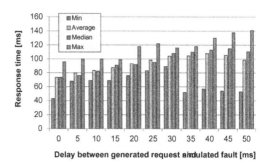

decided to simulate them after the requests with a delay of 5 ms multiplicity. In other words, the faults were simulated 0, 5, 10, 15 ms etc. after the request. The result of the experimental research are shown in Fig. 7. They are divided into several groups according to the delay between requests and simulated faults. For every group four values are given minimum, maximum, average, and median response time.

In the considered hot standby configuration, the faults of the active computing unit makes the remote I/O station switch the control source to the redundant unit. It was mentioned before that the fault is detected after three network cycles. As a result, the later the fault is simulated in relation to the request generation, the later the switchover routine is executed. Therefore, the maximum response time is increasing together with the fault simulation delay. On the other hand, it may happen that the fault is simulated so late, that the computing unit manages to send new output data (including the system response) to the remote I/O station before the unit is disconnected from the bus. It starts to happen when the delay is longer than 30 ms. The more the delay is increased, the more frequent are the responses before the fault simulation. It may be reasoned from the decreasing value of the average value of the response time.

Delays greater than 50 ms were not shown on the graph because with that delay the simulated faults stopped having any influence on the response time—the responses were always sent to the remote I/O stations by the computing unit before the faults were simulated.

The highest response time recorded during the experiment was 141 ms. That makes the cost of computing unit redundancy in transient state be as follows:

$$C_{RCUt} = \frac{max(T_{ResRCUt})}{T_{Res}} = \frac{141.0}{91.2} = 1.55 , \tag{4}$$

That means that during the computing units switchover the response time is greater by more than a half in comparison with the response time in system without redundancy. It should be noted that the measured maximum time is not the worst case. For example, would the automata cycle duration be longer, the redundancy costs could be even more significant.

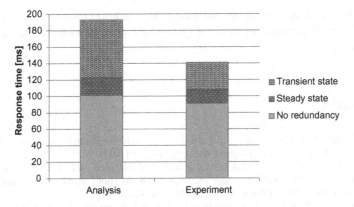

**Fig. 8** Comparison of the response time values between the theoretical analysis and the experimental results

## 4.4 Results Comparison

A comparison of the response time values between the theoretical analysis and the experimental results are shown in Fig. 8.

The results of the theoretical analysis are much greater than the values obtained during the experimental research as the analysis refers to the worst case.

## 5 Conclusions

Redundancy is used to increase the reliability of computer systems. It is often implemented during design of new systems, but also introduced into existing systems to make them invulnerable to faults of their most critical components. As the above summary of performed analysis and laboratory research shows, redundancy must be implemented with great care especially in the latter case. Otherwise, it may happen that the system is more reliable, but should not be exploited in order not to endanger human beings or environment to harm or damage caused by a missed deadline.

Another important thing is to distinguish two types of temporal costs of redundancy. Firstly, there are costs in the *steady state* of the system, which is associated with fault-free conditions. They make the system to response to requests with some delay in comparison to system without redundancy. Secondly, additional temporal costs are associated with fault occurrence of any of the redundant components. They worsen the response time even more. The latter case, called by the author as *transient state* of the system, is most crucial. The system is in steady state almost all of the time of its operation. Therefore, worse system responsiveness could be noted while running the system. The transient state on the other hand, occurs only occasionally but in most critical moments of the operation of the computer systems, i.e. after a fault of its redundant components and during the switchover routines. In that circumstances the temporal parameters of the system have great importance and any delay may be disastrous.

# References

1. Dinh, T.Q., Ahn, K.K., Marco, J.: A novel robust predictive control system over imperfect networks. IEEE Trans. Ind. Electron. **64**(2), 1751–1761 (2017)
2. Bauer, N.W., van Loon, S.J.L.M.B., van de Wouw, N., Heemels, W.P.M.H.M.: Exploring the boundaries of robust stability under uncertain communication: an ncs toolbox applied to a wireless control setup. IEEE Control Syst. **34**(4), 65–86 (2014)
3. Lu, Y., Nolte, T., Bate, I., Cucu-Grosjean, L.: A statistical response-time analysis of real-time embedded systems. In: 2012 IEEE 33rd Real-Time Systems Symposium, pp. 351–362, December 2012
4. Halawa, H.H., Hilal, Y.K., Aziz, G.H., Alfi, C.H., Daoud, R.M., Amer, H.H., Refaat, T.K., ElSayed, H.M.: Network fabric redundancy in ncs. In: Proceedings of the 2014 IEEE Emerging Technology and Factory Automation (ETFA), pp. 1–4, September 2014
5. Gaj, P., Jasperneite, J., Felser, M.: Computer communication within industrial distributed environment—a survey. IEEE Trans. Ind. Inform. **9**(1), 182–189 (2013)
6. Rzasa, W., Rzonca, D., Dariusz, A.S.: Analysis of challenge-response authentication in a networked control system. In: Conference: 19th International Conference on Computer Networks, Szczyrk, pp. 271–279 (2012)
7. Gupta, R.A., Chow, M.Y.: Networked control system: overview and research trends. IEEE Trans. Ind. Electron. **57**(7), 2527–2535 (2010)
8. Flak, J., Gaj, P., Tokarz, K., Widel, S., Ziebinski, A.: Remote monitoring of geological activity of inclined regions—the concept. In: Proceedings of Computer Networks—16th Conference, CN 2009, Wisła, Poland, 16–20 June 2009, pp. 292–301 (2009)
9. Jestratjew, A., Kwiecien, A.: Performance of http protocol in networked control systems. IEEE Trans. Ind. Inform. **9**(1), 271–276 (2013)
10. Cupek, R., Ziebinski, A., Franek, M.: Fpga based OPC UA embedded industrial data server implementation. J. Circuits Syst. Comput. **22**(8) (2013)
11. Zhou, Z., Tang, B., Xu, C.: Design of distributed industrial monitoring system based on virtual token ring. In: 2007 2nd IEEE Conference on Industrial Electronics and Applications, pp. 598–603, May 2007
12. Pedro, J., Burns, A.: Worst case response time analysis of hard real-time sporadic traffic in fip networks. In: Ninth Euromicro Workshop on Real-Time Systems, pp. 3–10 (1997)
13. Phan, R.C.W.: Authenticated modbus protocol for critical infrastructure protection. IEEE Trans. Power Deliv. **27**(3), 1687–1689 (2012)
14. Nguyen, V.Q., Jeon, J.W.: Ethercat network latency analysis. In: 2016 International Conference on Computing, Communication and Automation (ICCCA), pp. 432–436, April 2016
15. Cereia, M., Bertolotti, I.C., Scanzio, S.: Performance of a real-time ethercat master under linux. IEEE Trans. Ind. Inform. **7**(4), 679–687 (2011)
16. Wisniewski, L., Wendt, V., Jasperneite, J., Diedrich, C.: Scheduling of profinet irt communication in redundant network topologies. In: 2016 IEEE World Conference on Factory Communication Systems (WFCS), pp. 1–4, May 2016
17. Sestito, G.S., Turcato, A.C., Dias, A.L., Rocha, M.S., Brando, D., d. V. Torres, R.: Case of study of a profinet network using ring topology. In: 2016 IEEE International Symposium on Consumer Electronics (ISCE), pp. 91–96, September 2016
18. Cheminod, M., Durante, L., Valenzano, A.: Review of security issues in industrial networks. IEEE Trans. Ind. Inform. **9**(1), 277–293 (2013)

19. Gaj, P., Skrzewski, M., Stj, J., Flak, J.: Virtualization as a way to distribute pc-based function-alities. IEEE Trans. Ind. Inform. **11**(3), 763–770 (2015)
20. Natori, K., Oboe, R., Ohnishi, K.: Stability analysis and practical design procedure of time delayed control systems with communication disturbance observer. IEEE Trans. Ind. Inform. **4**(3), 185–197 (2008)
21. Jamro, M., Rzonca, D.: Measuring, Monitoring, and Analysis of Communication Transactions Performance in Distributed Control System, pp. 147–156. Springer International Publishing, Cham (2014)
22. Cucinotta, T., Mancina, A., Anastasi, G.F., Lipari, G., Mangeruca, L., Checcozzo, R., Rusina, F.: A real-time service-oriented architecture for industrial automation. IEEE Trans. Ind. Inform. **5**(3), 267–277 (2009)
23. Gaj, P.: Pessimistic useful efficiency of EPL network cycle. In: Proceedings of Computer Networks—17th Conference, CN 2010, Ustroń, Poland, 15–19 June 2010, pp. 297–305 (2010)
24. Widel, S., Flak, J., Gaj, P.: Interpretation of dual peak time signal measured in network systems. In: Proceedings of Computer Networks—17th Conference, CN 2010, Ustroń, Poland, June 15–19, 2010, pp. 141–152 (2010)
25. Gaj, P.: The concept of a multi-network approach for a dynamic distribution of application relationships. In: Proceedings of Computer Networks—18th Conference, CN 2011, Ustron, Poland, 14–18 June 2011, pp. 328–337 (2011)
26. Sun, F., Pan, X., Liu, Y., Chen, Y.: The analysis of stability of networked control systems based on industrial switched ethernet. In: 2008 7th World Congress on Intelligent Control and Automation, pp. 5682–5686, June 2008
27. Biswal, G.R., Maheshwari, R.P., Dewal, M.L.: System reliability and fault tree analysis of seshrs-based augmentation of hydrogen: dedicated for combined cycle power plants. IEEE Syst. J. **6**(4), 647–656 (2012)
28. Neves, F.G.R., Saotome, O.: Comparison between redundancy techniques for real time appli-cations. In: Fifth International Conference on Information Technology: New Generations (itng 2008), pp. 1299–1300, April 2008
29. Yu, X., Jiang, J.: Hybrid fault-tolerant flight control system design against partial actuator failures. IEEE Trans. Control Syst. Technol. **20**(4), 871–886 (2012)
30. Giorgetti, A., Cugini, F., Paolucci, F., Valcarenghi, L., Pistone, A., Castoldi, P.: Performance analysis of media redundancy protocol (mrp). IEEE Trans. Ind. Inform. **9**(1), 218–227 (2013)
31. Kim, M.H., Lee, S., Lee, K.C.: Kalman predictive redundancy system for fault tolerance of safety-critical systems. IEEE Trans. Ind. Inform. **6**(1), 46–53 (2010)
32. Vial, J., Virazel, A., Bosio, A., Girard, P., Landrault, C., Pravossoudovitch, S.: Is triple modular redundancy suitable for yield improvement? IET Comput. Digit. Tech. **3**(6), 581–592 (2009)
33. Thekilakkattil, A., Dobrin, R., Punnekkat, S., Aysan, H.: Optimizing the fault tolerance capa-bilities of distributed real-time systems. In: 2009 IEEE Conference on Emerging Technologies Factory Automation, pp. 1–4, September 2009
34. Restrepo-Calle, F., Martnez-lvarez, A., Guzmn-Miranday, H., Palomoy, F.R., Cuenca-Asensi, S.: Application-driven co-design of fault-tolerant industrial systems. In: 2010 IEEE Interna-tional Symposium on Industrial Electronics, pp. 2005–2010, July 2010
35. Gorawski, M., Marks, P.: Fault-tolerant distributed stream processing system. In: 17th Interna-tional Workshop on Database and Expert Systems Applications (DEXA 2006), 4–8 September 2006, Krakow, Poland, pp. 395–399 (2006)
36. Kwiecien, A., Sidzina, M., Mackowski, M.: The concept of using multi-protocol nodes in real-time distributed systems for increasing communication reliability. In: Proceedings of Computer Networks, 20th International Conference, CN 2013, Lwowek Slaski, Poland, 17–21 June 2013, pp. 177–188 (2013)
37. Zuloaga, A., Astarloa, A., Jimnez, J., Lzaro, J., Angel Araujo, J.: Cost-effective redundancy for ethernet train communications using hsr. In: 2014 IEEE 23rd International Symposium on Industrial Electronics (ISIE), pp. 1117–1122, June 2014
38. IEC: Industrial communication networks: Highavailability automation networks. Part 3: Par-allel Redundancy Protocol (PRP) and High Availability Seamless Redundancy (HSR) **IEC 62439-3** (2012)

39. Rentschler, M., Heine, H.: The parallel redundancy protocol for industrial ip networks. In: 2013 IEEE International Conference on Industrial Technology (ICIT), pp. 1404–1409, February 2013
40. Hoga, C.: Seamless communication redundancy of iec 62439. In: 2011 International Conference on Advanced Power System Automation and Protection, vol 1, pp. 489–494, October 2011
41. Automation, G.F.: Genius i/o, system and communications, November 1994
42. Stoj, J., Kwiecien B.: Real-time system node model—some research and analysis. In: Contemporary aspects of computer networks, vol. 1, pp. 231–239 Warsaw, Polnad, WKL (2008)

# SIPE: A Domain-Specific Language for Specifying Interactive Programming Exercises

**Jakub Swacha**

**Abstract** The paper introduces a new domain-specific language designed essentially for specifying exercises for interactive programming courses. The language covers the four crucial elements of interactive programming exercise, i.e.: metainformation, exercise description, solution checking and feedback, and aims at conciseness, human readability and ease of use. The paper discusses the issues with alternative forms of specifying programming exercises, presents the general concepts and syntax of the language, and reports on its implementation and validation.

**Keywords** Programming learning environments · Exercise specification · E-learning

## 1 Introduction

An important element of an effective teaching of computer programming is providing the students the ability to try the code they write as an exercise and obtain feedback on whether it is syntactically and semantically correct. Traditionally, it was a role of the instructor to check the solutions produced by the students, and provide them with an appropriate feedback. With the development of interactive programming learning environments, the instructor could be relieved from this time-consuming duty.

Automating this process brings also other major benefits, by allowing the students to proceed with their own pace, and get feedback whenever they fail or succeed, with no delay and regardless of the time of day. The value of automatic feedback for the effectiveness of programming learning has been confirmed in research [1].

J. Swacha (✉)
Institute of Information Technology in Management,
University of Szczecin, Szczecin, Poland
e-mail: jakubs@uoo.univ.szczecin.pl

© Springer International Publishing AG 2018
P. Kosiuczenko and L. Madeyski (eds.), *Towards a Synergistic Combination of Research and Practice in Software Engineering*, Studies in Computational Intelligence 733, DOI 10.1007/978-3-319-65208-5_2

15

It is crucial that the feedback, though automatic, should be meaningful, that is help the students understand the reasons of the errors they committed and possibly suggest ways to overcome them. This can only be achieved by triggering the feedback using specially designed tests based on the knowledge of the problem, appropriate solution methods, used programming language and the relevant typical programming mistakes [2].

Thus, the traditional instructor's procedure of exercise preparation consisting of three stages: conceive, explain to students, check solutions (of which the latter two have to be repeated for each lesson), is replaced by a new procedure, consisting of another three stages: conceive, describe, specify the tests and feedback (of which none has to be repeated for each lesson). Although there is a time gain from not having to repeat specifying the tests and feedback for each lesson, the difficulty and arduousness of the specification can make this gain disappear or even become a loss.

In this paper it is assumed that the way an interactive programming exercise is specified is related to the necessary instructor's effort. It can be argued that a well-designed language for describing programming exercises and specifying tests and feedback can reduce the time required to perform these steps both directly, by making it simpler to translate from concepts to specification and the result more concise, and indirectly, by making the specification simpler and more human-readable, thus easier to understand, correct, reuse, and improve. The aim of this paper is to introduce a domain-specific language designed essentially to meet these goals. For ease of reference, the language was called SIPE (from *a language for Specification of Interactive Programming Exercises*).

As much as most of the innovation comes from real-world needs, the inspiration for SIPE were the issues encountered by this author during many hours spent on specifying exercises for his own interactive programming learning environment [3]. These issues are discussed in Sect. 3 of this paper. Before that, Sect. 2 provides a short review of related work on programming exercise specification.

The key Sect. 4 describes the general design of SIPE and its syntax, whereas Sect. 5 focuses at the crucial ingredient of SIPE, that is the sublanguage used to specify the test and feedback parts of exercise specification. Section 6 describes the real-world implementation and validation of SIPE, and the final section concludes.

## 2   Related Work

Although interactive environments for programming learning have become a commodity these days, so far no standard for specifying exercises for such systems has been established and widely adopted, and often ad hoc-designed, proprietary formats are used for this purpose. The closest proposal to a what could be considered a standard is PExIL [4]. It covers all the primary aspects of exercise specification, including metadata and task description as well as definition of output tests and feedback. Still, PExIL is XML-based and as such does not meet the goals, presented above, of conciseness and ease of viewing and editing as text.

Although aimed at quite a different goal, but worth mentioning in this context, is the Unified Exercise Answering Format presented in [5], used as standardized form of reporting solutions of programming exercises.

Looking at the test and feedback specification alone, the usual approaches are to check the existence (or lack) of text patterns specified explicitly or using regular expressions. A general-purpose programming language is also often employed for this purpose, which may but need not be the language the solution is written in [2], but that seems to be a far too sophisticated tool in the case of interactive programming courses where the most of the performed tests are very simple. Even operating system shell scripts could be used for this purpose [6, p. 4], but such a tool, though relatively simpler, has a lot of limitations of its own.

Another approach is to use special annotation embedded in the model solutions to specify the feedback [7], yet it is not applicable, when there is no model solution.

The concept of providing a wrapper extending regular expressions was investigated in [8], though that solution uses a specific notation only for output- and input-checking, whereas mere regular expressions are used for pattern specification for source code checking.

## 3   Issues of Exercise Specification

The issues discussed below are based on actual issues encountered by this author during the development of his own interactive programming course [3]. Although the goal of this paper is not to document the way it was implemented (much of which can be found in [9]), some basic knowledge regarding it is necessary to understand the context of the issues.

The above-mentioned course uses JSON files to store information about the exercises. One such file contains specification of all exercises belonging to one course unit (lesson). The exercises are defined as an array of objects, each containing fields with the respective exercise specification elements, such as 'title' (text to display as the exercise title), 'task' (text of HTML description of what the student is instructed to do), 'outputHas' (array of search phrases defined with regular expressions each of which has to match the contents of the output generated by the student's solution for it to be considered correct), or 'outputHasHint' (corresponding array of text messages containing feedback for the student in case search phrases were not matched, e.g. if the first search phrase was not matched, the first message is displayed as hint).

The explicit examples of some of the discussed issues are listed in Table 1 at the end of this section, which also presents how such issues disappear when using SIPE.

**Table 1** Test pattern specification issues and how they are solved using SIPE

| Issue | RegEx-based example | SIPE equivalent |
|---|---|---|
| Optional space | `/\(\s*limit\s*=\s*50\s*\)/` | `'( limit = 50 )'` |
| Special char's | `/\*\*|f\.a\s*\*f\.a/` | `any('**', 'f.a * f.a')` |
| Escaped char's | `/1\.\.\.\\n\\t2\.\.\.\\n/` | `'1...\n\t2...\n'` |
| Valid code | `/^(?:[^#"']*(?:'(?:[^']|\\')*'|` <br> `"(?:[^"]|\\")*"))*[^#"']*\bfor\b/m` <br> (for Python; does not even cover multi-line strings and escaped backslashes) | `for` |
| Within block | `/if\s+x\s*<\s*0\s*:(?:\s*|\n(\s+)` <br> `(?:.*\n\1))x\s*=\s*-\s*x/` | `'x=-x'` <br> in compound `'if x<0'` |
| Exact number of occurrences | `/^(?:[^i]|i[^f]|\Bif|if\B)*` <br> `\bif\b(?:[^i]|i[^f]|\Bif|if\B)*$/` | `just 1 if` |
| Alternatives without specified order | `/\(\s*(?:(?:a_1\D+)?1\D+\d` <br> `\D+(\d)[\s\S]+\(\s*(?:(?:a_1\D+)?` <br> `1\D+\1|\D+[23])|\D+([23])\D+(?:a_1` <br> `\D+)?(\d)\D+(\d)[\s\S]+\(\D*(?:a_1` <br> `\D+)?(?:\2\D+\4|(?:\3|\4)))/` | `distinct 2 each (` <br> `any ('( 1','a_1 = 1'),` <br> `2, 3) in bracket '('` |
| Ordered lists | `/^\D*(?:0\D+)*(?:1\D+)*(?:2\D+)*` <br> `(?:3\D+)*(?:4\D+)*(?:5\D+)*` <br> `(?:6\D+)*(?:7\D+)*(?:8\D+)*` <br> `(?:9\D+)*(?:10\D+)*$/` | `incr /\b1?\d\b/` |
| Pattern references | `/[a-zA-Z_][a-zA-Z0-9_]*\.a\s*\*` <br> `\s*[a-zA-Z_][a-zA-Z0-9_]+\.b|` <br> `[a-zA-Z_][a-zA-Z0-9_]+\.b\s*\*` <br> `\s*[a-zA-Z_][a-zA-Z0-9_]+\.a/` | `[a-zA-Z_][a-zA-Z0-9_]*` <br> `-> $name` <br> `any(` <br> `$name$'.a *` <br> `'$name$'.b',` <br> `$name$'.b *` <br> `'$name$'.a')` |

The examples are real-world and come from [3]

## 3.1 Exercise Description

The description of an interactive programming exercise should contain all the information necessary for the student to develop an acceptable solution for the exercise. It consists of three elements, containing respectively:

- text introducing the student to the problem area covered by the exercise, and possibly explaining language constructs or algorithm design techniques that will be needed to solve the exercise and that he/she may not know;
- task specifying the expected outcome of running the student's solution program, and possibly the additional requirements (e.g. the language constructs or functions that the student should or should not use);
- initial state of the solution source code that the student is expected to develop or correct (may be none).

The first issue experienced with all the three elements mentioned above is the required use of delimiters and separators ensuing from the fact that JSON has to conform to JavaScript syntax rules. The text has to be put into quotes, and while it lasts for several editor lines, a backslash has to be appended at the end of each non-terminal line. The most frequent reason of syntax errors with exercise specification found was forgetting to put the '\' at the line end. The second: missing a comma between elements of an array—usually after appending a new element to it in a new line.

The second issue was the existence of forbidden characters and phrases, and the required use of escape characters as a work-around for this limitation. As for the HTML-formatted text, the most frequent issue was caused by comparison operators ('<', '>'), which were often forgotten to be replaced with respective entities ('&lt;', '&gt;'). As for the source code (both the initial solution source code and examples contained in the exercise description), the problem was with quotes and the backslash. This issue was augmented by the fact these characters have to be escaped both in the language of the source code (Python in this instance) and the language of the specification (JavaScript) which not only made it easy to make a mistake, but also made source code strings containing a number of literal backslash characters (which thus were required to be doubled twice) to be represented in a form highly obscure to a human reader.

The third issue was the frustrating nuisance due to the verbosity of HTML markup, requiring typing tag names within triangle brackets, and doing it twice—at the begin (sometimes also with class attributes) and the end of every marked phrase.

## 3.2   Test Specification

Using regular expressions to define the patterns that should or should not be found in the solution source code or the output of its execution revealed a number of issues, ranging from small but frequent nuisances (like the optional whitespace problem) to rare but serious difficulties requiring use of sophisticated and elaborate expressions (like the multiple compound expressions with different order of items problem). While all of them reduce the human readability of the specification, the latter make it even more difficult to analyze, verify and reuse. In the list below, nine such problematic cases are described.

- The optional whitespace problem. Most programming languages permit optional whitespaces, which have to be accommodated for in the test patterns. As a result, the single spaces are replaced with whitespace wildcards, obfuscating the pattern specification.
- The special characters problem. The special characters of regular expressions (e.g. '.', '(', ')', '[' or '*' are very frequently used in program source, thus also frequently included in patterns, in which they have to be escaped. As a result, many backslashes are inserted, obfuscating the pattern specification.

- The escaped characters problem. Backslash is used as the escape character in regular expressions and many programming languages, where it can also serve other purposes. As a result, the number of inserted backslashes grows.
- The valid code problem. Most of the patterns looked for in the solution source code is expected to match valid program instructions, but they can as well match literal string constants and comments. A special pattern can be inserted in front of the actual pattern sought to avoid that but it obfuscates the pattern specification.
- The code within block problem. In languages forming blocks by indentation (such as Python), looking for pattern that is expected inside a compound statement defined by another pattern requires a sophisticated regular expression, using group references. Similar problem is with looking for patterns within string literals, which can be expressed using various notations (different delimiters, joined single-line, multi-line).
- Exact number of occurrences problem. Regular expressions checking non-occurrence of patterns look awkward and contribute to low readability of test specification.
- Alternatives without specified order problem. Sometimes there are several ways to do something and there is need to check that more than one way of doing it was actually used in the solution. An exemplary exercise included checking whether a student is able to call functions using different order of three parameters. The used regular expressions was long and complex.
- Ordered list problem. Exercises may require the solution to produce output in sorted order. Checking that using regular expressions is a nuisance, especially if the range of expected values is large.
- Pattern reference problem. The regular expression dialect implemented in JavaScript is somewhat limited, for instance it lacks references to user-defined patterns (references to match groups are of course supported). In case specific pattern is used multiple times in a regular expression, its definition has to be repeated each time, making the expression unnecessarily long and thus difficult to grasp.

Table 1 presents examples of the issues described above; its rightmost column shows how the same goal can be achieved using far less complicated code written in SIPE.

## 4  SIPE: Basic Concepts and Syntax

The main design aims behind SIPE were:

- make the exercise specification easy to write in a general-purpose text editor,
- make all the necessary functionality (e.g. adding metainformation, special formatting of phrases within exercise description, specifying check patterns) available to the user,

- avoid the verbosity and complexity issues experienced with HTML, JSON and solution checks based only on regular expressions,
- keep it simple, especially the often used specification constructs,
- make it flexible, as individual course authors may have needs that would be hard to foresee.

In a quest for simplicity, a flat document structure was chosen as the basis for SIPE. This means that there are no logical brackets (like e.g. '{'and '}' in JSON), and only the position of an element in a document defines its relation to other elements. Thus, if element type *lesson* (see further down for more details) describes a set of exercises, *exercise*—a single exercise, and *initcode*—a specific component of an exercise, the content of an *initcode* element belongs to the last preceding *exercise*, and the content of an *exercise* element belongs to the last preceding *lesson*. Thanks to that, whole SIPE document is a sequence of just two interleaving types of blocks:

- section header, defining the meaning of the content of a subsequent section,
- section content, i.e. the actual specification of the exercise element denoted by the header.

## 4.1  Section Header

The section header is defined as follows (from now on, ABNF notation [10] is used for specification of key SIPE language elements):

```
section-header = section-name [ "." content-type ] ":"
                 *WSP [ section-title ] LF
```

The predefined section names are listed in Table 2. Only the *exercise* and *task* sections are required (without them, the document could not be called an exercise specification). As for the others, for the sake of flexibility, SIPE takes on a 'defined if exists' approach: the remaining sections are optional (they can be omitted), but:

- if a section is included in a document, it should contain exactly what the specification defines for such a section,
- if the content defined by the specification is included in a document, it should be put into the section having relevant name.

In other words, the predefined section names cannot be used for purposes other than specified, and their corresponding content types cannot be kept in sections having other names. The parser accepts any section name conforming to requirements (only letters are allowed). Note that all the structural elements of SIPE (including section names) are case-insensitive.

**Table 2** Predefined SIPE section names

| Section name | Content | Default content type |
|---|---|---|
| Lesson | Exercise collection title and metadata | Name: Value |
| Exercise | Exercise title and metadata (required!) | Name: Value |
| Intro | Text introducing the concepts needed to solve the task | Markdown |
| Task | Text explaining the task the student is supposed to perform (required!) | Markdown |
| Precode | Source code that will be injected at the beginning of the student's code before its execution | Plain text |
| Initcode | Source code that will be presented to the student as a skeleton for the solution | Plain text |
| Postcode | Source code that will be injected at the end of the student's code before its execution | Plain text |
| Inputparameters | Comma-separated values that should be fed to the tested program (exercise solution) as input | Plain text |
| Checksource | Code to check the tested program source code | SCSL |
| Checkinput | Code to check the tested program user-specified input | SCSL |
| Checkoutput | Code to check the tested program output | SCSL |
| Checkerrors | Code to check error messages generated by the tested program | SCSL |

The content type can be changed from default by specifying the actual content type following the section name. The predefined names for content types are: *HTML*, *MD* (for Markdown), *NameValue* (for Name:Value pairs), *plain* (for plain text) and *SCSL* (for the Solution Check Specification Language). The parser accepts any content type name conforming to requirements (only letters are allowed). The interpreter may but cannot be guaranteed to handle user-defined content types.

The section title is optional, and, in practice, is used only for *lesson* and *exercise* sections. The section header ends with the end of line.

## 4.2 Section Content

The section content is defined as follows:

```
section-content = content LF LF labelchar
```

*Content* can be a sequence of any characters excluding the ending sequence (two line ends and the *labelchar* character). Thanks to this simple notation, no delimiters or escape characters need to be introduced to the actual content (as is the case with

XML or JSON), regardless of its type. The only limitation is that the content cannot contain the ending sequence, which is hardly an issue in practice.

The default *labelchar* is colon (':'). It can hardly be expected in natural language texts and most programming languages to be found at the line beginning following an empty line. However there are cases where such awkward notation could be encountered—for instance, consider a course on batch programming, where a colon at the line beginning defines a label. For this reason, the *labelchar* can be redefined to any other character or even a sequence of characters using one of the predefined options (*endcontent*) specified in *lesson* metadata. The list of predefined metadata for *lesson* and *exercise* section contents is given in Table 3.

The two sections containing exercise description that is to be presented to the student (*intro* and *task*) must allow the instructor to include text having various formatting (standard text, text with emphasis, headings, programming instructions within sentences, blocked code examples) as well as illustrations or links to other resources. Although HTML provides all this functionality, its markup is verbose and there is number of other issues (see Sect. 3 of this paper) that make it tiring for hand editing. For this reason, Markdown [11] has been chosen as a default content type for the *intro* and *task* sections. It provides all the functions mentioned above

**Table 3** Predefined metadata types

| Metadata name | Content |
| --- | --- |
| Lesson section | |
| Language | Natural language used to teach the lesson (required!) |
| Programminglanguage | Programming language taught in the lesson (required!) |
| Endcontent | Character (or their sequence) marking the end of section content |
| Author | Author(s) of the content |
| Email | E-mail of the corresponding author(s) |
| Version | Version of the lesson |
| Created | Date of creating the first version of the lesson |
| Edited | Date of last edit of the current version of the lesson |
| Exercise section | |
| id | Exercise identifier. By default, exercises are numbered incrementally from 1 in their order of specification |
| Difficulty | Exercise difficulty level |
| Reward | Points or other virtual reward for completing the exercise |
| Precodevisible | Whether *precode* is visible to the student. By default, it is invisible |
| Postcodevisible | Whether *postcode* is visible to the student. By default, it is invisible |
| Blocked | Whether exercise can be taken by the student without completing others first. By default, it is |
| Unblocks | List of exercises that will be unblocked after completing this one |

with an ease of marking down text by inserting but few characters instead of repetitive HTML tags.

The three sections containing source code (*precode*, *initcode* and *postcode*) as well as predefined input parameter values (*inputparameters*) do not require any formatting (as they are displayed in a code editor with a visual scheme of its own, or not displayed at all), and are represented in plain text which means there is no need for delimiters or character escape codes.

The remaining four predefined sections (*checksource*, *checkinput*, *checkoutput*, *checkerrors*) describe tests that the submitted exercise solution will have to be passed through primarily to determine its correctness, but also to generate hints that will help the student to solve the problems causing syntactical, execution or semantic errors, as well as improve possibly even a correct but suboptimal solution. Being aware of the issues with the use of mere regular expressions, a new, specially designed language (i.e., a sublanguage of SIPE) has been created for this purpose. The Solution Check Specification Language (SCSL), as it was called, is described in the following section (with more information provided in a forthcoming publication [12]).

Due to article length limitations it is impossible to present here a real-world SIPE-based exercise specification. However, to give a glimpse of how it looks like, a very simple exercise specification expressed in valid SIPE is provided below:

```
lesson: First things first

:exercise: Hello World!

:intro:
print command prints

:task:
Print "Hello World!"

:initcode:
"Hello World!"

:checksource:
req print => You missed the command.

:checkoutput:
req "Hello World!" => I cannot see "Hello World!".

:
```

## 5  Test and Feedback Specification

In SIPE, the automatic tests of the exercise solution, as well as the feedback generated after passing (or failing to pass) these tests, are specified, by default, using Solution Check Specification Language (SCSL). Each set of tests (i.e., the content of a *checksource* section of a given exercise) is represented as a sequence of statements which are executed from top to down. Statement is defined as follows:

```
statement = [boolvar *wsp "=" *wsp]
            ["if" 1*wsp boolexp sep] [test 1*wsp]
            phrase [sep "=>" sep feedback] *wsp
            [%d13] %d10 [%d13] %d10
sep = 1*wsp / *wsp [%d13] %d10 *wsp
```

Observe that:

- the result of a match can be stored in a boolean variable,
- a match can be executed on a condition defined with a specified boolean expression which can reference any boolean variable set earlier (this allows for, e.g., generating hints depending on a number of factors),
- the *test* can be either *req* or *forbid*, following the well-established practice of using both pessimistic ("the solution is wrong unless a certain structure is present") and optimistic ("the solution is correct unless a undesirable structure is present") rules [13, p. 22],
- *test* is optional, when it is not specified, the result of the pattern matching does not affect the solution correctness check, but the specified feedback can still be generated (if the pattern does not match; useful for improvement hints) and the result can be stored in a boolean variable (thus obtaining component result for a composite check),
- *feedback* is a possibly multiline text generated if the *phrase* matches (for *forbid*) or does not match (otherwise);
- the statement ends with double linefeed (i.e., an empty line); this allows for splitting long check specifications into multiple lines as well as increases the readability.

The actual pattern (or a sequence of patterns) to be sought is given in *phrase* element, consisting of one or several (separated with commas) patterns. If several patterns are given, they are considered by default as sequence, i.e. each pattern is matched from the position at the end of the match of the pattern preceding it. The following operators can be used to modify this behavior:

- *each*: match all the elements in any order,
- *none*: match none of the elements,
- *any*: match at least one of the elements,
- *select* num: match exactly *num* of the elements in any order, e.g.:
- `req select 1 (a, b, c)` reports correct only if there is either *a* or *b* or *c* matched (but not a combination of these).

By default, the pattern is matched in the whole text (as defined by the SIPE section name; moreover, for *checksource* the match can only start at valid code, i.e. not inside comments or literal string constants). The match range can be limited to the block defined after *position* operator, which defines the relation between the sought pattern and its context (*block*), and can be:

- *in*: the pattern will be matched inside the *block* (see below for details),
- *after*: the pattern will be matched no sooner than the block ends,
- *follows*: the pattern will be matched right after the block ends (only language-specific whitespaces are allowed in-between).

The *block* can be specified either by type only (the pattern will be matched in each block of that type) or also specified by *phrase* (the pattern will be matched only in those blocks of that type, which match such *phrase*). There are five types of *block*:

- *bracket*: matches within the (specified) kind of brackets (one of '()', '< >', '[]', '{}'); in *checksource* section, brackets outside of language-specific valid code (e.g. inside comments or literal string constants) are ignored; the context *phrase* may include text preceding the opening bracket, but it must explicitly include the opening bracket and it cannot include any other brackets; the sought pattern may include the opening and closing brackets (as anchors),
- *line*: matches within the (specified) line (in *checksource* section, defined by language-specific boundaries, e.g. '\' may disable line end in some languages);
- *compound*: only applicable to *checksource* section, matches in the (specified) compound statement, having language-specific boundaries; this is designed especially for languages which do not use brackets (like '{}' in C) to delimit compound statements, as e.g. Python; the context *phrase* matches from the beginning of the compound statement (i.e. including its header);
- *string*: matches in the (specified) string (delimited with quotes; in *checksource* section, defined by language-specific boundaries, e.g. '\"' may disable string end in some languages);
- *comment*: only applicable to *checksource* section, matches in the (specified) code comment.

For patterns returning multiple matches, their required quantity can be set using the following operators:

- *just* num: the pattern must match exactly *num* times; for instance, the statement req just 3 /\b\d\d\b/ reports correct only if the number of occurences of two-digit numbers is 3;
- *atleast* num: the pattern must match *num* or more times,
- *atmost* num: the pattern must match *num* or less times,
- *between* num1 *and* num2: the pattern must match at least *num*1 and at most *num*2 times,
- *multiply* num: the pattern must match $num*x$ times, where $x$ is any integer $\geq 1$; for instance, the statement req multiply 2 /\b\d\d\b/ reports correct only if the number of occurences of two-digit numbers is even.

Also, additional requirements can be defined regarding the order of matched values:

- *distinct*: none of the fragments matched by the pattern can repeat
- (e.g. `req distinct /\d +/` reports correct only if there is at least one number, and, if there is more than one of them, they all have different values),
- *same*: each fragment matched by the pattern must have the same content,
- *incr*: each subsequent fragment matched by the pattern must have greater value (lexicographic order is used for non-decimals);
- *decr*: each subsequent fragment matched by the pattern must have smaller value (lexicographic order is used for non-decimals); for instance, the statement
- `req decr /\b\d +\b/` reports correct only if there is at least one number and, if there is more than one of them, they are ordered with decreasing value.

There are several ways of expressing a pattern available:

- *word* (e.g.: `print`): a sequence of non-space characters (from a limited set, mostly alphanumeric), intended to match single instruction names and identifiers; there is no need for delimiters other than space; strings identical to SCSL keywords cannot be specified this way;
- *words* (e.g.: `'x/2'`): a sequence of characters delimited with apostrophes; it has special properties: a space matches any whitespace sequence and a double quotation mark matches any language-defined literal string constant delimiter;
- *string* (e.g.: `"Good Bye"`): a sequence of characters delimited with double quotation marks; they match literally the given string (no wildcards, no special properties);
- *regex* (e.g.: `/p[a-z]*t/`): a regular expression (JavaScript flavor) delimited with slashes;
- *varref*: a name of variable preceded with '$' character, it is an alias of a pattern defined earlier using the '->' operator (see below);
- *varvalue*: a name of variable preceded with '#' character, it matches the last value matched by the referenced pattern or returns no match if there was no prior match to set the value.

A single pattern can be represented using a combination of the ways listed above:

```
/[A-Za-z_][0-9A-Za-z_]*/' = '#val1
```

Each pattern can be labelled for later reuse using -> operator, e.g.:

```
/-?\b\d +\.\d +\b/ - > float
```

Table 1 in Sect. 3 gives more examples of SCSL expressions.

## 6   Implementation and Validation

The SIPE (including SCSL) has been defined in Augmented Backus-Naur Form (ABNF) [10]. The SIPE section parser and interpreter has been implemented directly in JavaScript by this author. Micromarkdown.js [14] is used as the Markdown converter. The JavaScript parser for SCSL has been generated automatically using APG [15] and served as the basis for its interpreter. Currently, only an interpreter tuned for Python source code (i.e. the exercise solutions are expected to be written in Python) has been developed, still it is configurable and can be easily adapted to any other programming language by replacing a set of callback functions.

The interpreter is being used as an engine for version 2.0 of the Python interactive course by this author [3]. All its code will be released as open-source as soon as the new course is published. At the moment of writing these words, the course is in the process of content conversion, improvement and extension which includes translation to SIPE of almost two hundred exercises and over one thousand tests.

The SIPE and its interpreter were validated using multiple exercises from the mentioned course [3]. This included all exercises from the initial part of the course and hand-picked exercises featuring tests based on especially complex regular expressions from later parts of the course. No major issues were encountered, and minor issues were used as a base for improvements in language specification and its interpreter.

## 7   Conclusion

In the paper, a new domain-specific language has been described, designed for specifying exercises for interactive programming courses. The language defines a simple structure of programming exercise specification, including metadata, exercise description, solution checking and feedback. It aims at conciseness, human readability and ease of use, especially the ability to edit the specification in generic text editors without nuisances.

For this reason, Markdown [11] has been chosen as the tool for exercise description whereas for the purpose of test and feedback specification, a new sublanguage, SCSL has been proposed [12]. It allows for a very concise definition of patterns typically sought for in exercise tests, and the resulting specification is far more readable than one based merely on regular expressions.

The language has been implemented in JavaScript and validated on numerous exercise examples. It will be soon published as an engine behind a new version of this author's interactive Python course [3].

# References

1. Fernandez, J.L.: Automated assessment in a programming tools course. IEEE Trans. Edu. **54** (4), 576–581 (2011)
2. Swacha, J.: Scripting environments of gamified learning management systems for programming education. In: Peixoto de Queirós, R.A, Teixeira Pinto, M. (eds.) Gamification-Based E-Learning Strategies for Computer Programming Education, pp. 278–294. Information Science Reference, Hershey, PA, USA (2017)
3. Swacha, J.: Interaktywny kurs języka Python. http://uoo.univ.szczecin.pl/~jakubs/kurs (2016)
4. Queirós, R., Leal, J.P.: Making Programming Exercises Interoperable with PExIL. In: Ramalho, J.C., Simões, A., Queirós, R. (eds.) Innovations in XML Applications and Metadata Management: Advancing Technologies, pp. 38–56. IGI, Hershey, PA, USA (2013)
5. Zeng, C., Xie, L., Chen, G., Arikawa, S., Ishihara, Y.: OPECSS: an on-line programming exercise checking support system. In: Proceedings of Conference on Educational Uses of Information and Communication Technologies, pp. 199–205. Publishing House of Electronics Industry, Beijing, China (2000)
6. Joy, M., Griffiths, N., Boyatt, R.: The BOSS online submission and assessment system. JERIC **5**(3), art. 2 (2005)
7. Gerdes, A., Jeuring, J., Heeren, B.: An interactive functional programming tutor. In: Proceedings of the 17th ACM Annual Conference on Innovation and Technology in Computer Science Education, pp. 250–255. ACM, New York, NY, USA (2012)
8. Hadiwijaya, R.I., Liem, M.M.I.: A domain-specific language for automatic generation of checkers. In: International Conference on Data and Software Engineering, pp. 7–12. IEEE, Yogyakarta, Indonesia (2015)
9. Swacha, J.: Implementacja interaktywnego kursu programowania w technologii webowej. Studia Inform. Pomerania **3**(41), 103–112 (2016)
10. Crocker, D., Overell, P. (eds.): Augmented BNF for Syntax Specifications: ABNF. RFC 5234. Network Working Group. https://tools.ietf.org/html/rfc5234 (2008)
11. Gruber, J.: Markdown. http://daringfireball.net/projects/markdown (2004)
12. Swacha, J.: Exercise Solution Check Specification Language for Interactive Programming Learning Environments. Forthcoming (2017)
13. Goedicke, M., Striewe, M., Balz, M.: Computer aided assessments and programming exercises with JACK, ICB-Research Report, No. 28. http://hdl.handle.net/10419/58160 (2008)
14. Waldherr, S: Micromarkdown.js. http://simonwaldherr.github.io/micromarkdown.js (2017)
15. Thomas, L.D.: APG ... ABNF Parser Generator. http://www.coasttocoastresearch.com (2017)

# Managing Software Complexity with Power-Generics

Stan Jarzabek

**Abstract** Complexity of software quickly approaches the limits of what today's programming paradigm can handle. Similarities (i.e., similar requirements, design solutions, as well as program structures) are inherent in software domain. In the paper, we discuss unexploited potentials of software similarities to ease management of complex software systems. We describe the concept of *power-generics* to exploit this potential. A key idea is meta-program level flexible parameterization, without restrictions of C++ templates or Java generics. To illustrate the concept, we discuss ART (Adaptive Reuse Technique) that extends conventional programming paradigms with an unconventional generative technique, in a synergistic and easy to adopt way. With ART, we illustrate general concepts discussed in the first part of the paper.

**Keywords** Generic design · Software clones · Software reuse · Maintenance · Generative techniques · Meta-programming

## 1 Introduction

Software systems can comprise 10s of millions LOC, with thousands of inter-related components, reaching the level of complexity that becomes difficult to handle with today's technology. We'll be surely challenged by even larger and more complex software in the future. Ultra-Large Scale System initiative [16] targets systems of systems, comprising billions lines of code.

This study was supported by grant S/WI/2/2013 from Bialystok University of Technology and founded from the resources for research Ministry of Science and Higher Education.

S. Jarzabek (✉)
Faculty of Computer Science, Bialystok University of Technology, Białystok, Poland
e-mail: s.jarzabek@pb.edu.pl

© Springer International Publishing AG 2018                                         31
P. Kosiuczenko and L. Madeyski (eds.), *Towards a Synergistic Combination of Research and Practice in Software Engineering*, Studies in Computational Intelligence 733, DOI 10.1007/978-3-319-65208-5_3

Control over software of that scale is unthinkable if developers continue to be exposed to software complexity[1] proportional to the software size. Especially software maintenance, which is almost exclusively done at the level of code, exposes developers to such complexity. Not surprisingly, up to 80% of software costs go to maintenance. In development of new software, Domain-Specific Languages (DSL) supported by generators and software reuse are the two paradigms that allow us to escape software complexity proportional to the software size. A notable example of DSL/generator paradigm is a compiler-compiler that allows developers to work with grammatical specifications, orders of magnitudes smaller and simpler than the generated compiler code. Software reuse reduces the size of code that needs be developed proportionally to the size of reused components, although we must also add the complexity of interfaces between reused components and the rest of a software system. In an established approach to software reuse called Software Product Lines (SPL) [5], generic code components are built for a specific domain, and domain-specific SPL architecture facilitates component interfacing and integration. However, these two above mentioned technologies provide solutions only in a narrow scope of well-understood and stable application domains.

In this paper, we are interested in a general-purpose method to reduce the cognitive complexity of programs that would be independent of an application domain in the way Java generics or C++ templates are. We particularly consider long-lived software systems, or systems of systems, where evolution leads to many component versions that are embedded in many system releases in various configurations. It is a major challenge to evolve system releases in independent directions, while keeping in sight what they all have in common and how they differ. This challenge has been only partially mastered by today's technologies such as configuration management systems.

Decomposition (componentization), abstraction, separation of concerns and generic design are the main principles that help us manage the complexity of software systems. In this paper, we focus on generic design, but in Sect. 8 we discuss also other principles and their contributions as well as limitations to manage software complexity.

By *similarity patterns* we mean similar program structures, of significant importance, repeated many times in variant forms within a system or across systems, and their successive versions (releases).

Many software systems are densely populated by similar program structures of all kinds and granularity, such as code fragments (known as *software clones* [12]),

---

[1]System complexity if often measured in terms of the number of conceptual elements and relationships among them that must be understood in order to understand the system. It is this kind of cognitive complexity that we mean in the paper.

code components (similar functions, classes, source files, directories, etc.) [8], recurring configurations of components (known as *structural clones* [2]), recurring patterns of collaborating components, and repetitions on a subsystem scale.

> By unifying these repetitions with powerful generic representations, we could contract the number of conceptual elements (components and their interactions) in a program solution space, reducing the cognitive complexity of the subject software system. We use term *power-generics* to mean any technique capable of such unification, without compromising maintainability or other important program qualities.

STL [11], a library of C++ data structure classes, is a classic example of simplifications that can be achieved by avoiding repetitions. However, type-parameterization (called templates in C++) and other techniques such as iterators successful for generic data structures and algorithms are not sufficient to achieve genericity in other domains [8, 9, 17, 21] (even in STL some of the repetitions remain unresolved [1]).

Power-generics—unrestricted generics—must rely on templates that provide a more flexible way of parameterizing software components than conventional parametrization techniques. We illustrate the concept with a generative technique of ART (Adaptive Reuse Technique). ART templates can represent generically any group of similar program structures. Of course, not all similarities are worth considering for unification, and it is up to a designer to decide unification of which similarities will result in reducing program complexity. Conventional programs, automatically obtained from ART templates, contain repetitions that must necessarily be there, but they do not contribute to the program complexities as perceived by a programmer anymore.

The paper is organized as follows: Sects. 2–4 discuss the background and motivate power-generics as means to manage software complexity. In Sects. 5 and 6, we describe implementation of power-generics concepts in ART. Section 7 discusses further the nature of software similarities. In Sect. 8, we review major approaches to managing software complexity, and when appropriate contrast them with power-generics in ART. Conclusions end the paper.

## 2 Good Clones, Bad Clones

While practitioners are aware of much repetitions in software, they also know how difficult it is to avoid them. Problems with implementing effective reuse strategies [6] evidence these difficulties, as well. Similarities are sometimes obvious, sometimes implicit and dispersed across system components and, therefore, difficult to spot. The differences among similar program structures are often irregular and arbitrary. Our studies suggest that repetitions can comprise large parts of software systems, and avoiding them with conventional techniques is often either impossible or not achievable without compromising other important design goals [9, 17, 21].

With the exception of poor design and careless reuse via copy-paste-modify, repetitions are most often intentional, created in a good cause, have their own merits and purpose [13]. For example, by cloning code developers may achieve better software performance, higher reliability, easier portability or other goals. Some repetitions may be there because of limitations of a programming technology in its ability to define effective generic design solutions. Such repetitions cannot be avoided given design goals and technology used [1, 9, 10, 15, 21]. Yet other repetitions may be induced by a technology, for example, may occur as the result of pattern-driven development in modern component platforms such as.NET™ or JEE™, bringing beneficial uniformity of how problems are solved and standardization of software organization.

Whatever the reason for repetitions, the very presence of them is an evidence that the software uniqueness space is smaller than its physical size. If noticed and tackled in a generic way, software similarities open rich possibilities for program simplification, easier maintenance and reuse rates surpassing what we can achieve with today's component-based approaches.

In our studies of new, well-designed programs, we typically find 50–90% of code contained in similar program structures, repeated many times. For example, the extent of the redundant code in the Java Buffer library was 68% [9], in parts of STL (C++)—over 50% [1], in J2EE Web Portals—61% [21], and in certain ASP Web portal modules—up to 90% [17].

## 3  Power-Generics—What's the Deal?

How to benefit from similarities, given that most of the repetitions cannot be eliminated from programs? We believe the very nature of this dilemma requires us to look beyond conventional programming techniques, trying to address the problem at the program meta-level. In that way, we can keep clones in an executable program intact, but represent them non-redundantly at the meta-program level to simplify the cognitive complexity of a program (i.e., make it easier to comprehend and maintain for a developer). This is exactly what power-generics discussed in this paper are meant to achieve.

Here is why and how meta-level non-redundancy may reduce cognitive complexity of software:

Suppose we have 10 user interface forms $a_1$, ..., $a_{10}$ (e.g., data entry forms). Each form, say $a_i$, interacts with five business logic functions $b_i$, $c_i$, $d_i$, $e_i$, $f_i$ (e.g., data validation rules or actions to be performed upon data entry at certain fields). If each user interface form and each business logic function is implemented as a separate component, we have to manage $10 + 10 * 5 = 60$ components and 50 interactions. This situation is shown in Fig. 1a.

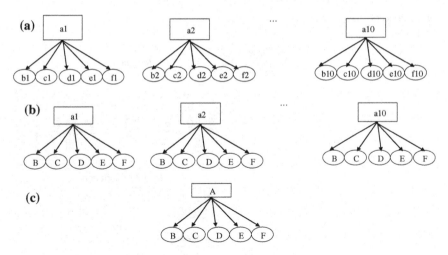

**Fig. 1** The impact of unifying similarities patterns on complexity

Now let us assume that there is considerable similarity among the 10 user interface forms $a_i$, and also among business logic functions in each of the groups $b_i$, $c_i$, $d_i$, $e_i$, $f_i$ ($i = 1, \ldots, 10$). Suppose we represent 10 similar business logic functions $b_i$ with a generic function B, and likewise groups of functions $c_i$, $d_i$, $e_i$, $f_i$ ($i = 1, \ldots,$ 10) with generic functions C, D, E, F, respectively. We reduce the solution space to $10 + 5 = 15$ components and 50 interactions, with the slightly added complexity of a generic representation. The interactions have become easier to understand, as each form interacts with components that have been merged into five groups rather than 50 distinct components. This situation is shown in Fig. 1b.

Suppose we further represent 10 user interface forms $a_i$ with one generic form A. We reduce the solution space to $1 + 5 = 6$ components, with five groups of interactions, each group consisting of 10 specific interactions, plus the complexity of a generic representation. This situation is shown in Fig. 1c.

Software validation and changes performed at the level of the generic components would be easier than at the level of concrete components due to a smaller number of distinct components and interfaces that have to be analysed and clearer visibility of the impact of changes. Generic software structures also form natural units of reuse, potential building blocks for other similar systems (e.g., forming a Software Product Line [5]).

In the above discussion, we presented in abstract terms what we found in many applications of our experimentation. One such application was a Project Collaboration Environment (PCE), a web portal supporting teams in project development. Figure 2 shows a PCE architecture, with functional modules at the top.

PCE supports communication among project team members. For that, PCE maintains information about staff, projects, who works on which projects, tasks assigned to various staff members, and other information that matters in collaboration. For each entity such as Staff, Project, Task, Product or File there is a

**Fig. 2** Project collaboration
web portal

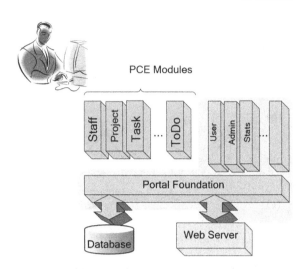

corresponding PCE module to manage it. For example, module Staff allows a PCE
user to *Create, Edit, Delete, Display, Find* or *Copy* data about staff members, assign
staff members to projects, etc. Similar operations apply to other PCE modules.

PCE modules are deployed on top of the Portal Foundation that implements
common services reused by the portal modules. PCE also reuses some of the
standard modules such as User, Admin, Stats, History and others. We do not
discuss the Portal Foundation or the implementation language/platform for PCE any
further as they do not matter for our argumentation here.

Each operation such as *CreateStaff, CreateProject or CreatTask* is implemented
by components from user interface (UI), business logic (BL) and database
(DB) layers, shown as boxes (Fig. 3). Each component contains several classes.
User interface classes implement various UI forms to display or enter data; business
logic classes implement data validation and other actions specific to various
operations and/or entities; database classes define data access and table structures.
Arrows indicate runtime interactions among classes contained in respective components via message passing.

Same shading of boxes indicates high similarity between classes in the corresponding components. Operations in a group *CreateStaff, CreateProject, Create-
Task*, etc., are implemented by similar program structures formed by patterns of
collaborating components. Each group of such similar program structures forms
so-called structural clones [2].

## 4 Difficulties to Build Generic Solutions

So why it is difficult to represent similar patterns of collaborating components of
Fig. 3 with a generic pattern shown in Fig. 4?

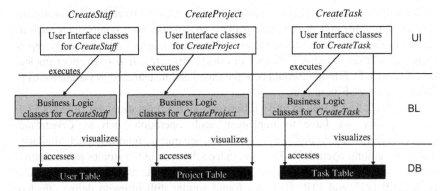

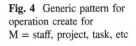

**Fig. 3** Similar patterns of collaborating PCE components

**Fig. 4** Generic pattern for operation create for M = staff, project, task, etc

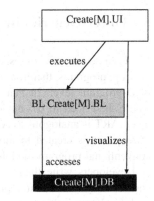

There are a number of problems that hinder forming generic operations with conventional techniques such as type parameterization, inheritance or design patterns:

*First*, generic representation of groups of similar methods or classes may be already a problem. Despite much similarity, classes implementing operations differ in arbitrary details, most of which are not of type-parametric nature. For example, entities Staff, Project or Task have different attributes and data entry validation rules. Creating Staff may require extra functionality that is not applicable to Project or Task, and vice versa. These differences in entities' semantics trigger differences among classes implementing *CreateStaff*, *CreateProject*, *CreateTask* and other operations in the group. Similar classes in respective components may differ in some algorithmic details of method implementation or details of method signature; some classes may have extra methods or different attribute declarations as compared to other classes in the same group. These problems complicate forming generic classes, which is a prerequisite to forming generic components *Create* [M]. UI, *Create* [M].BL, and *Create* [M].DB.

*Second*, even if we succeed in unifying similar classes, there is no conventional syntactic program unit to represent patterns of collaborating components such as shown in Fig. 3. Generic components and patterns can be only implicitly represented in software system's file and directory structures. It is even more unclear how a generic PCE module could be represented and instantiated in variant forms to build specific PCE modules.

If we can't find a way to unify differences among operations for different PCE modules, then we have to implement each operation such as *CreateStaff*, *CreateProject*, ..., *EditStaff*, *EditProject*, etc. separately, ignoring much similarity that exists among operations. This was indeed the case in lab studies and industrial projects where PCE was designed using a number of technologies such as ASP [17], J2EE [21], and PHP [18]. We found similar difficulties to define effective generic design solutions even in STL [1].

## 5  Power-Generics in ART

As discussed in previous sections, duplications are often unavoidable and intentional in programs, therefore it is best to realize the concept of power-generics at the meta-program level. In that way we can keep clones in an executable program, where then need to be, but represent them non-redundantly at the meta-program level. ART is analogous to Aspect-Oriented Programming [14], where a *meta-level extra plane* is created to modularize concerns that otherwise inevitably crosscut conventional program modules (classes).

Developers write their program solutions in programming languages of their choice. They use ART to represent generically groups of similar program structures such as shown in Fig. 3.

In a nutshell, ART (1) represents each significant group of similar program structures as templates, (2) delineates the differences among specific program structures in each group as deltas from their generic form, (3) records the exact location of each such structure in a program, and (4) automates deriving specific structures from ART templates to produce a proper program that can be compiled.

Figure 5 shows an outline of ART representation of PCE in PHP/ART. Boxes are ART templates organized into a hierarchy, with lower-level, small-granularity generic structures being building blocks for higher-level, and larger-granularity ones.

At each level, we unify groups of similar programs structures as follows: Level 6 unifies similar methods, Level 5—similar classes, Level 4—similar components in user interface, business logic and database layers, Level 3—patterns of components implementing operations such as *CreatStaff*, *CreateProject*, *CreateTask*, and Level 2—PCE modules. **SPC** (Level 1) contains specifications of customizations required to derive PCE modules from **PCE-template** and templates below it. Template **PCEtemplate** contains specifications of customizations required to derive operations. An arrow between two templates: X → Y reads as "X adapts Y", meaning

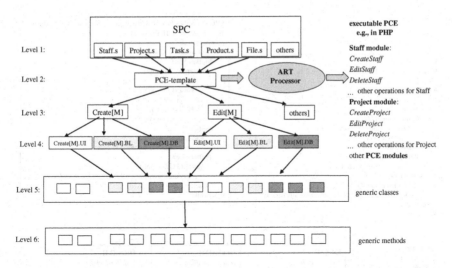

**Fig. 5** Hierarchical unification of similarity patterns

that X controls adaptation of Y. Custom operations for specific entities (e.g., *CreateStaff*) are obtained by adapting respective generic operations (e.g., Create [M]).

Figure 6 shows the details of PCE code organized into ART templates and instrumented with ART commands. ART Processor interprets templates starting from **SPC**, traverses templates along **#adapt** links, adapting them and emitting PCE code in PHP. Customizations specific to a given operation (*Create* or *Edit*) for a given module (Staff or Project) are defined under relevant **#option**s in **#select**. Yet other, more ad hoc customizations are addressed with **#insert** into **#break**. ART Processor propagates variable values as well as **#insert**s down to all the **#adapt-**ed templates, overriding any default values assigned to variables in **#adapt-**ed templates that cater for common cases. For example, customizations for module Staff affect operation *Create* in a number of ways shown in Fig. 6.

ART variables and expressions provide template parameterization mechanisms and help exercise navigation during template customization process. #**set** command assigns a list of values to a variable. ART rules for propagating variable values and **#insert**s ensure that common behaviour becomes customized for specific operations, without affecting other operations. These rules allow variables to coordinate chains of customizations for a given source of change, spanning multiple templates.

The **#insert** into **#break** command has a similar effect to weaving aspect code in AOP [14]. Functions specific to the operation Create are defined in the frame For_Staff_Only. With **#insert** In_Create> in the SPC, we insert Create-specific functions at designated variation points <**break** In_Create> in templates PCE-template and Create[M].

Each iteration of **#while** loops in **SPC** and in **PCE-template** derives code for one operation such as *CreateStaff* or *EditProject*. The I'th iteration of the loop uses

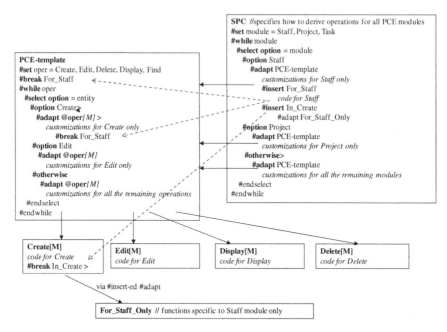

**Fig. 6** Flexible "composition with adaptation"

the i'th value of its control variable **oper** listed in the respective **#set** command. Unique customizations required for specific operations are specified under a suitable option of the **#select** command.

By varying specifications, we can instantiate the same collection of templates in different ways, deriving different programs from it.

In ART representation, commonalities are cleanly separated from specifics: Generic templates (such as **PCE-template** or **Create[M]** in our example) define what is common to a given group of similar structures. Upper level templates define specifics—customizations relevant to specific instances of a template. Such separation is one of the key guiding principles of ART design, leading to a reusable and maintainable software representation. The way ART Processor propagates customizations across templates helps us to achieve this separation, and to make templates reusable in multiple contexts. ART variable values and **#insert**s propagate across templates from top to the bottom, via **#adapt** chains. Each **#adapt** chain may define different customizations. ART variables mark and control multiple customization points related to the same source (e.g., customization for a specific buffer class or a specific PCE module or operation).

# 6 ART Trade-Offs

The benefit of non-redundancy achieved with ART power-generics does not come for free. Designing generic solutions is always a challenge which requires more talent and skill than building a concrete program. Power-generics require an up-front investment in exchange for future system design, development, and evolution benefits.

By separating genericity issues from the core constructs typically supported by programming languages, we can address genericity concerns without compromising program runtime properties. But as we move towards less restrictive parameterization mechanisms, we also decrease type-safety of parameterized program solutions.

Industrial applications have demonstrated that ART is easy and fast to learn, and its benefits usually greatly outweigh the cost of the added learning curve [17]. At the same time, the return on investment may be quick and substantial. Industrial applications have also contributed to our understanding of added complexities and their offsets such as the difficulty to encompass the two intermixed levels, in base programming language(s) and ART. However, we must keep in mind that ART templates contain much useful information that help developers in software maintenance and reuse, in addition to complete information about the subject program(s) itself. Debugging is yet another area into which ART induces extra complexities. On the other hand, most errors can be traced to the least tested code, which is usually concentrated only in top-level templates. ART Processor tags each generated line with the SPC line that was active when the line was generated. Thus when an error occurs at some random line of code, developers know where to look to find its most likely cause.

# 7 Software Similarity Phenomenon

Just like certain computational aspects necessarily crosscut conventional program modules [14, 20], certain similarity patterns lead to software redundancies. These redundancies show as similar program structures of different type and granularity spreading through software systems.

Design-level similarities often represent domain-specific abstractions. They manifest themselves as patterns of classes, components or subsystems. Representing design-level similarities in a generic form is particularly useful, as it explicates the knowledge of how domain concepts are realized in the solution space. Furthermore, generic representation can be reused within a given system, or across similar systems.

Conventional methods—component-based, architecture-centric approaches as well as language-level features such as generics—often fail to reap benefits of software similarities, for variety of reasons. Many repetitions are just meant to be there, so the very idea of eliminating them is not acceptable. Repetitions introduced for performance or reliability reasons, and those induced by pattern-driven development (e.g., on.NET or JEE) belong to this category.

In cases where there is no higher-level reason for repetitions, limitations of a programming language or platform often hinder conventional generic solutions. Ad hoc, irregular variations among similar program structures do not make it easy to refractor repetitions into a generic representation. Software design is constrained by multiple, sometimes competing, design goals, such as development cost, usability, simplicity, performance, or reliability. A generic solution to unify similarity patterns can only win if it is natural and synergistic with all the other design goals that matter.

Yet other similarity patterns may not correspond to any conventional abstraction (such as class or component) suitable for their representation. Patterns of collaborating classes or components are often in this category.

ART is not constrained by rules of the underlying programming language or platform. Therefore, any recurring program structure worth attention can be captured in a generic form independently of its type, granularity or the reason for its presence.

Looking for deeper roots of the software similarity phenomenon, feature combinations are one of the major forces triggering the spread of similar program structures within and across systems. Functional abstractions (e.g., components, classes or functions/methods) implement the net effect of the required feature combination that is necessary for correct execution. As we combine features, variations arise that force us either to find a unifying generic solution, or to create multiple program structures in variant forms. In industrial Software Product Line projects, feature combinations may lead to thousands of similar component versions [6].

## 8  Related Work

In this section, we discuss Brooks's perspectives on software complexity [4], and software engineering concepts and advances that help us manage software complexity.

Thirty years ago, Brooks divided difficulties faced by software technology into essential and accidental, where essential difficulties are "the difficulties inherent in the nature of the software—and accidents—those difficulties that today attend its production but that are not inherent."[2] Accidental difficulties can be helped with technological advances, but essential ones—cannot. Brooks argues that software complexity is an inherent, essential property of software. It cannot be reduced below a certain threshold with methods and tools, therefore "building software will

---

[2]All quotes in this section refer to [4].

always be hard. There is inherently no silver bullet." Until today, none of the technological advances has proven Brooks wrong.

Brooks explains software complexity as follows: as software grows in size, we see "an increase in the number of different elements. In most cases, the elements interact with each other in some nonlinear fashion, and the complexity of the whole increases much more than linearly." The notion of software complexity we use in this paper (footnote at the first page of this paper and discussion in Sect. 3) coincides with the above quoted definition.

"The complexity of software is an essential property, not an accidental one. Hence, descriptions of a software entity that abstract away its complexity often abstract away its essence." Here, fine differentiation of notions is required to avoid confusion: Should we consider a BNF grammar specification of a programming language a "description" of a parser program in the above sense? After all, a complier writer needs only write BNF grammar specifications of a programming language, and a parser generator automatically generates thousands of code lines that build up an optimized parser for that language. The complexity of BNF grammar specifications is low, while the complexity of the parser code is high and irreducible at the code level. However, to understand the parser code it is not enough to study BNF specifications. One also has to look at the parser generator. In his seminal paper [4], Brooks has not discussed generation-based solutions such as compiler-complier. It is therefore the author's opinion that Brooks's remark on irreducible complexity refers to software descriptions that include a complete information about an operational program solution, not specifications in a DSL (such as BNF) that require a domain-specific generator to produce fully operational program.

How about a non-redundant power-generics software representation in ART discussed in this paper? Not only does program representation in ART contain complete information about an operational program, but it also contains additional traceability information showing the impact of requirements on various code components. At the same time, the complexity of a power-generics representation is lower than that of the actual program derived from that representation. This refers to both complexity measured in number of elements and their relationships, and cognitive complexity from the developer's point of view. It is not clear to the author how to interpret Brooks's remark on irreducibility of essential software complexity in case of software representations such as power-generics.

The history of attempts to master the complexity of software, both the product and the development process, is as long as the history of computing. Invention of high-level programming languages was a major step towards improving software productivity. Modularization, abstraction, separation of concerns, genericity, and software reuse have been prime principles to tackle software complexity. However, there are no easy solutions.

*Componentization of software* ("divide and conquer") into modules has been always a fundamental approach to tackle software complexity. Software components tend to be coupled one with another, in explicit (via interfaces) and implicit ways. Understanding one component requires knowledge of yet other components.

Layered design with highly cohesive and loosely coupled modules, with carefully defined module interfaces and information hiding helps us arrive at a possibly simple modular structure. However, difficulties to realize component-based reuse hint at fundamental limits of what we can achieve by means of conventional "divide and conquer" componentization. Lego-like componentization fits well physical artefacts that traditional engineering deals with. However, it does not reflect the nature of software equally well [3]. Software architectures are more enigmatic and less stable than hardware architectures. Software components are also less stable, and interact in more complex ways than hardware components. They are often changed, rather than replaced, therefore must be much more flexible than hardware components. Software components have not become a commodity to the extent as optimistic forecasts of late 1990s were predicting (e.g., vide Cutter Consortium and Gartner reports).

However, the most serious limitation of componentization in fighting complexity is that as long as we develop, test and maintain software in terms of concrete, executable components, we are exposed to the complexity proportional to system size. In a system of thousands of interacting components, the complexity of validating, testing and controlling maintenance changes is bound to explode beyond control. In author's view, this complexity is part of the Brook's essential complexity [4]. Only when coupled with some form of software reuse, manual or automated by generators, componentization opens options for more substantial complexity reduction.

*Separation of concerns* is a powerful concept, however it is not easy to bring it down from the concept to the design and implementation levels. Only a restricted number of computational concerns can be componentized, or separated with techniques such as AOP [14] or MDSC [20]. We refer the reader to [7] for detailed discussion of difficulties to bring separation of concerns from concept to the code level. While separation of concerns remains a powerful method to manage complexity at the concept level, its ability to manage complexity at the code level is limited.

For long *software reuse* has been considered as means for reducing software complexity and achieving radical improvements in software productivity. The prime objective for reuse is the same as for generic design—to avoid repetitions, i.e., to avoid writing similar code all over again and maintaining it. The current state-of-the-art in reuse is based on architecture-centric, component-based and Product Line concept [5]. Having scoped variability (usually modelled as feature diagrams), a Software Product Line architecture (SPLA) is designed to accommodate variant features. With component-based design, genericity of a SPLA is achieved by stabilizing component interfaces and localizing the impact of variant features to possibly small number of components. However, for variant features that have crosscutting effect on SPLA components, this goal cannot be easily achieved. A component affected by variant features explodes into numerous versions, similarities among component versions cannot be easily spotted, and reuse opportunities offered by such similarities are usually missed. Given thousands of variant features and complex features inter-dependencies arising in industrial Software Product

Lines, the cost of finding "best matching" component configurations for reuse, and then the follow up component customization, integration and validation, may become prohibitive for effective reuse [6].

With ART, we treat similarity patterns induced by variant features as first class citizens, no matter whether they have a crosscutting effect or not. Power-generics unify similarity patterns at any granularity level and of any type—from a subsystem, to a pattern of components, to a component, to a class and to a program statement within a class implementation. For that reason, the application of ART often extends the scope and rates of reuse achievable by means of conventional techniques.

Conventional component-based reuse is most effective when combined with architecture-centric, patter-driven development which is now supported by the major platforms such as.NET™ and J2EE™. Patterns lead to beneficial standardization of program solutions and are basic means to achieve reuse of common service components. IDEs support application of major patterns, or developers use manual copy-paste-modify to apply yet other patterns. ART representation can enhance the benefits of modern platforms by automating pattern application, and emphasizing the visibility of patterns in code. Pattern-driven design facilitates reuse of middleware service components, but tends to scatter application domain-specific code. With ART, we can package and isolate otherwise scattered domain-specific code into reusable generic components. Such extensions improve development and maintenance productivity, and allow reuse to penetrate application business logic and user interface system areas, not only middleware service component layers.

*Abstraction* can shield developers from some of the essential complexity. Abstraction can play more than merely descriptive role in narrow, well-understood domains, whereby we can make assumptions about the problem domain semantics. By encoding domain knowledge into generators, we let developers work at abstract levels of program description in a Domain-Specific Language (DSL), with a generator filling the missing details. This can lead to substantial productivity gains, best exemplified by compiler-compilers. DSLs and generators are an automated, therefore most productive implementation of software reuse concepts.

Both non-redundant software representation in ART and DSL/generator solutions abstract away cognitive software complexity, but there is one important difference: The non-redundant meta-program still contains complete details of the actual program it represents. This is in contrast with DSL/generator solutions, where specifications in DSL can be very simple (e.g., BNF specifications of a programming language), but to understand the actual program generated from specifications (a parser) we need to look into the generator (parser generator). This becomes an obstacle in situations when a developer needs to change or enhance the generated code, a problem we discuss in more details below.

Advancements in modelling and generation techniques led to Model-Driven Engineering (MDE) [19] where multiple, inter-related models are used to express domain-specific abstractions. Models are used for analysis, validation (via model checking), and code generation. Platforms such as Microsoft Visual Studio™ and

Eclipse™ support generation of source code using domain-specific diagrammatic notations.

While helpful and insightful, generator-based solutions face problems that have hindered their deeper penetration of the programming practice. The first problem is the likelihood of disconnecting models from code during evolution. This occurs when the DSL cannot cater for unexpected evolutionary changes and developers modify the generated code. Round-trip engineering has been notoriously difficult to achieve, and modified code needs to be maintained by hand, with no help from models or a generator. Such disconnection is most likely to happen if multiple and independently evolving program versions originate from a generator: A change specific to one such program, if implemented in the generator, automatically propagates to other programs that may not need the change—a mostly undesirable effect. Implementing a change into a specific program disconnects it from the generator. Supporting multiple component versions and system releases via generators poses a fundamental problem.

The above problem is compounded by the difficulty of integrating multiple models/generators to build systems we need. Typically, a problem domain served by a generator covers only a part of a given system. Strategies for integrating multiple domain-specific generators and embedding them into systems implemented using yet other techniques have yet to be developed. One of the reason for success of compiler generators is that compilation on its own is a self-contained domain.

The first mentioned problem motivated the invention of Frame Technology™ [3] at Netron Inc. ART is based on Frame Technology™, and comments below derived from Frame Technology™ experiences apply to ART. Netron Inc. used DSLs from which conventional generators produced code contained in templates (such as shown in Fig. 6), so that programmers could place all custom code in SPCs, separated from the generated code. In that way, customization and re-generation of code could be repeated independently of each other. Frame Technology™, viewed as a generation engine, is free of the model-code disconnection problem, as developers perform all the maintenance/reuse work at the level of templates, never changing code produced from templates.

Frame Technology™ addresses also the second mentioned problem of integration. Not only do templates support multiple component versions and system releases via generators, but they can also integrate multiple models/generators when building or evolving complex systems. Netron's customers routinely build and evolve multimillion line systems this way [3].

## 9 Conclusions

Parameterization and other generic design techniques, as well as advances in software reuse only partially exploit the potential of software similarities to manage software complexity. There is much evidence that software clones—recurring similar program structures—of all kinds, big and small spread across software systems. In many cases, these repetitions are intentional, so clones cannot be simply removed from programs. Nevertheless, clones contribute much to the cognitive complexity of programs. Today systems have grown to millions lines of code and systems of systems that are envisioned for the future will be bigger. We proposed power-generics approach to tackling similarity patterns in software as a possible way to better manage software complexity. In the paper, we discussed the proposed approach in the context of other efforts to manage software complexity, and illustrated the approach with meta-level technique ART that provides power-generics mechanisms.

**Acknowledgements** Thanks are due to Mr. Paul Bassett, the inventor of Frame Technology™ and a cofounder of Netron, Inc, for his generous contributions to our projects on ART and XVCL, and for his insightful comments on this paper.

Author is thankful to anonymous reviewers for insightful comments that helped him a lot to improve the paper.

# References

1. Basit. H.A., Rajapakse, D.C., Jarzabek, S.: Beyond templates: a study of clones in the STL and some general implications. In: International Conference Software Engineering, ICSE'05, St. Louis, USA, May 2005, pp. 451–459
2. Basit, H.A., Jarzabek, S.: Detecting higher-level similarity patterns in programs. In: ESEC-FSE'05, European Software Engineering Conference and ACM SIGSOFT Symposium on the Foundations of Software Engineering, ACM Press, September 2005, Lisbon, pp. 156–165
3. Bassett, P.: Framing Software Reuse—Lessons From Real World. Yourdon Press, Prentice Hall (1997)
4. Brooks, F.P.: No silver bullet, essence and accidents of software engineering. Comput. Mag. **20**(4), 10–19 (1987)
5. Clements, P., Northrop, L.: Software Product Lines: Practices and Patterns. Addison Wesley (2002)
6. Deelstra, S., Sinnema, M., Bosch, J.: Experiences in software product families: problems and issues during product derivation. In: Proceedings of Software Product Lines Conference, SPLC3, Boston, Augest 2004, pp. 165–182
7. Jarzabek, S., Kumar, K.: On Interplay between Separation of Concerns and Genericity Principles: beyond code weaving. Comsis J. Comput. Sci. Inf. Syst. **13**(3), 731–758 (2016); also in Kosiuczenko, P., Smialek, M. (eds.): Proceedings of 17th KKIO Software Engineering Conference, Miedzyzdroje, September 2015, Chapter 8, pp. 119–136, (Best Paper award)
8. Kumar, K., Jarzabek, S., Dan, D.: Managing big clones to ease evolution: linux kernel example. In: Federated Conference on Computer Science and Information Systems, FedCSIS, 36th IEEE Software Engineering Workshop, September 2016, pp. 1767–1776
9. Jarzabek, S., Li, S.: Eliminating redundancies with a "composition with adaptation" meta-programming technique. In: Proceedings of ESEC-FSE'03, European Software Engineering Conference and ACM SIGSOFT Symposium on the Foundations of Software Engineering, September 2003, Helsinki, pp. 237–246
10. Jarzabek, S.: Effective Software Maintenance and Evolution: Reused-based Approach. CRC Press Taylor and Francis, (2007)

11. Home page of SGI STL. http://www.sgi.com/tech/stl/
12. Kamiya, Toshihiro, Kusumoto, Shinji, Inoue, Katsuro: CCFinder: a multilinguistic token-based code clone detection system for large scale source code. IEEE Trans. Soft. Eng. **28**(7), 654–670 (2002)
13. Kasper, C., Godfrey, M.: Cloning considered harmful' considered harmful. In: Proceedings of 13th Working Conference on Reverse Engineering, October 2006, Benevento, pp. 19–28
14. Kiczales, G.J., Lamping, A., Mendhekar, A.C., Maeda, C., Lopes, C., Loingtier, J.-M., Irwin, J.: Aspect-oriented programming. In: European Conference on Object-Oriented Programming, Finland, Springer, LNCS 1241, pp. 220–242 (1997)
15. Kumar, K., Jarzabek, S., Dan, D.: Managing big clones to ease evolution: linux kernel example. In: Federated Conference on Computer Science and Information Systems, FedCSIS, 36th IEEE Software Engineering Workshop, September 2016, pp. 1767–1776
16. Ultra-Large-Scale Systems. http://www.sei.cmu.edu/uls/; also Northrop, L.: Ultra-Large Scale Systems: The Software Challenge of the Future, Software Engineering Institute, June 2006, ISBN 0-978656-0-7
17. Pettersson, U., Jarzabek, S.: Industrial experience with building a web portal product line using a lightweight, reactive approach. In: ESEC-FSE'05, European Software Engineering Conference and ACM SIGSOFT Symposium on the Foundations of Software Engineering. ACM Press, September 2005, Lisbon, pp. 326–335
18. Rajapakse, D.C., Jarzabek, S.: Using server pages to unify clones in web applications: a trade-off analysis. In: International Conference on Software Engineering, ICSE'07, Minneapolis, USA, May 2007
19. Schmidt, D.: Model-Driven Engineering, pp. 25–31. Computer, IEEE (2006)
20. Tarr, P., Ossher, H., Harrison, W., Sutton, S.: N degrees of separation: multi-dimensional separation of concerns. In: Proceedings of International Conference on Software Engineering, ICSE'99, Los Angeles, pp. 107–119 (1999)
21. Yang, J., Jarzabek, S.: Applying a generative technique for enhanced reuse on J2EE platform. In: 4th International Conference on Generative Programming and Component Engineering, GPCE'05, 29 September–1 October 2005, pp. 237–255

# A Prototype Tool for Semantic Validation of UML Class Diagrams with the Use of Domain Ontologies Expressed in OWL 2

**Małgorzata Sadowska**

**Abstract** Existing domain ontologies are a source of domain knowledge about the specific areas of interest. The article presents a prototype tool for semantic validation of UML class diagrams. A knowledge base for validating the diagram is the domain ontology expressed in OWL 2 selected by the user of the tool. The tool automatically recognises if the diagram is compliant, not contradictory or contradictory to the selected domain ontology. Additionally, it proposes a set of suggestions what should be corrected in the class diagram in order for the diagram to more adequately reflect the needed domain.

**Keywords** UML · OWL 2 · Semantic validation

## 1 Introduction

OWL 2 Web Ontology Language (OWL 2) [1] is a knowledge representation language for defining ontologies. The OWL ontologies, which satisfy syntactic conditions listed in [2] (in Sect. 3), are called OWL 2 DL ontologies and have semantics expressed in *SROIQ* description logic [1]. *SROIQ* was designed to provide many additions to OWL–DL in order to offer a satisfactory complexity [3, 4]. What is important, for practicability of reasoning it is still decidable. In the following description OWL always means OWL 2 if not stated differently.

Domain ontologies are expected to provide a knowledge base about specific application areas, therefore they have to be consistent. The previous work [5], outlined an idea of a method for automatic semantic validation of Unified Modeling Language (UML) class diagrams [6] with the use of domain ontologies expressed in OWL 2. This paper is focused on describing the tool which implements the validation method.

M. Sadowska (✉)
Institute of Informatics, Wroclaw University of Science and Technology,
Wroclaw, Poland
e-mail: m.sadowska@pwr.edu.pl

© Springer International Publishing AG 2018                                      49
P. Kosiuczenko and L. Madeyski (eds.), *Towards a Synergistic Combination of Research and Practice in Software Engineering*, Studies in Computational Intelligence 733, DOI 10.1007/978-3-319-65208-5_4

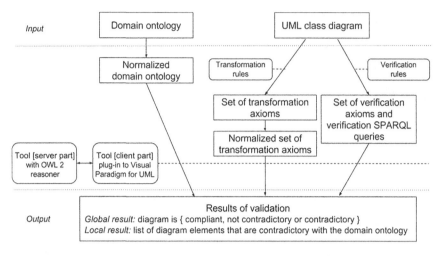

**Fig. 1** The revised illustration of the method of semantic validation of UML class diagrams

The proposed method and the tool are intended to address a practical problem of software engineering how a modeller can be sure that the developed UML class diagram is semantically correct. Using ontologies allows to conduct the validation process without the necessity of having the expertise provided by domain experts, which is usually expensive and time consuming. In the proposed approach UML class diagrams are confronted with the OWL ontologies which serve as a knowledge base about the domain.

The method of semantic validation of UML class diagrams, outlined in [5], was extended with additional steps which are presented in Fig. 1. There are two input elements of the method: a domain ontology selected by the user of the tool and a UML class diagram. The necessary preliminary requirement before the method can be applied is that the UML class diagram and the domain ontology must follow one agreed domain vocabulary (Fig. 2).

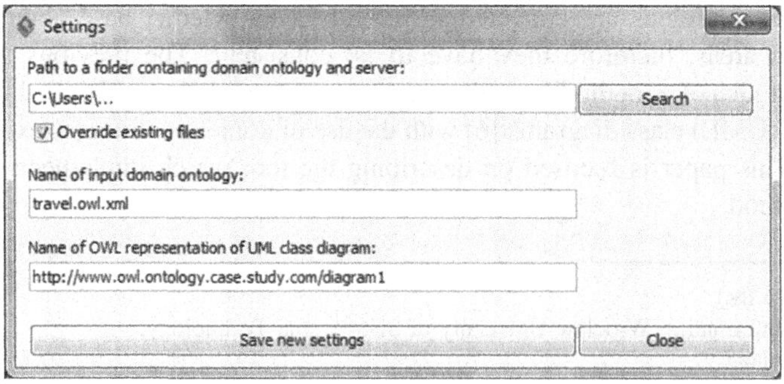

**Fig. 2** The settings form

The validation method requires a transformation of the UML class diagram to its OWL 2 representation. The transformation is double track, and as motivated in [5] required introducing transformation and verification rules in order to correctly and fully capture the semantics of the subsequential diagram elements. In the revised method, the verification axioms are additionally supported by verification SPARQL [7] queries. The role of verification axioms and verification SPARQL queries has not changed—both are intended to assure that the reengineering transformation (from the ontology to the diagram) would not remain in conflict with the semantics of UML class diagram. Due to the limited space, this article is not intended to present the details of the transformation, which is extensive. The reader can refer to a number of publications which present selected aspects of OWL to UML mapping, e.g. [8–14].

The method takes into consideration all static elements of UML class diagrams, which are important from the point of view of pragmatics. The tool checks the compliance with the domain ontology of UML classes, attributes of classes (with the multiplicity), associations (with the multiplicity of the association ends), generalizations between classes and between associations, generalization sets, structured datatypes and enumerations.

The new element in the method is the normalization of the domain ontology (and later in the process of validation—the normalization of the set of transformation axioms). Section 3 illustrates an example of semantically equivalent OWL 2 DL ontologies. With the use of the normalization it is easy to algorithmically compare two ontologies with the unified vocabulary, even if they are built of textually different axioms. Such a comparison is crucial in the validation method. The full process of normalization introduces rules aimed to refactor all subsequent OWL 2 constructs. The normalization only changes the syntax but does not affect the semantics of the input OWL 2 ontologies. The current research is focused on defining a full list of normalization rules for OWL 2 DL ontologies.

The correctly defined OWL 2 DL domain ontology allows to perform the needed reasoning which is conducted in the tool. More specifically, the tool checks the ontology consistency [15] with the use of the selected inference engine—HermiT.[1] The normalized domain ontology is iteratively modified by adding axioms from the normalized transformational part of OWL representation of UML class diagram. Each new axiom added to the normalized domain ontology entails a risk of making the ontology inconsistent, therefore the consistency check is conducted in every iteration of the algorithm. Finally, the last iteration of the algorithm presents the results of the validation.

The remaining part of this article is organized as follows. Section 2 describes related works. Section 3 illustrates an example of semantically equivalent ontologies. Sections 4 and 5 present architecture and features of the tool with an example of diagram validation. Section 6 outlines limitations of the validation method and

---

[1]HermiT OWL Reasoner—website: http://www.hermit-reasoner.com/.

the tool. Section 7 proposes some ontology-based suggestions for correcting the diagram. Finally, Sect. 8 concludes the paper.

## 2 Related Work

For the best knowledge of the author, no tool allows for semantic validation of UML class diagrams with the use of domain ontologies expressed in OWL 2. The presented tool is aimed to contribute to this field.

There are several tools for visualizing OWL ontologies. For example, OWL2UML [16] is a Protégé plug–in, which automatically generates UML diagram for an active OWL ontology with the use of OMG's Ontology Definition Metamodel. ProtégéVOWL [17] (also a Protégé plug-in) for visualization of OWL ontologies is based on Visual Notation for OWL Ontologies [18]. OWLGrEd [19] is a UML style graphical editor for OWL, in which the UML class diagram notation is extend with Manchester-like syntax for the missing OWL features. There are also tools for visualizing other then OWL ontological formalities, e.g. in [20], a tool for creating UML class diagram from SUMO ontology is proposed. The tool described in this paper also contains a supportive feature to visualize a selected part of the OWL domain ontology in the form of UML class diagram. This feature is useful for capturing the most important entities which should be presented on the diagram because in the validation method both the diagram and the ontology must follow one agreed vocabulary. Nonetheless, visualizing of OWL ontology is not used in the method of diagram validation itself.

## 3 Example of Semantically Equivalent Ontologies

The semantics of 1–3 example domain ontologies are the same, even though the ontologies contain different axioms. The normalization process is aimed to bring the ontologies written with the use of various OWL constructs to the same form which can be easily compared without the need of transforming axioms to the constructions in description logic. The presented OWL 2 constructs are written with the use of functional-style syntax [2].

Let us consider an OWL domain ontology which only contains one axiom:

**Domain ontology 1** *DisjointUnion(:Child:Boy:Girl)*

The *DisjointUnion(C CE$_1$ CE$_2$)* [2] axiom states that a class *C* (here::*Child*) is a disjoint union of the class expressions *CE$_1$* and *CE$_2$* (here::*Boy* and:*Girl*), all of which are pairwise disjoint. Following specification of OWL 2 [2], *DisjointUnion* axiom can be seen as a syntactic shortcut for the following two axioms:

**Domain ontology 2** *EquivalentClasses(:Child ObjectUnionOf (:Boy:Girl))*
*DisjointClasses(:Boy:Girl)*

Following definitions of OWL 2 constructs (Sect. 13.2 of [2]), one could modify the axiom further, even if it will not change the semantics. For example, OWL offers a class expression *ObjectComplementOf(CE)* [2], which contains all individuals that are not instances of the class expression *CE*. Double use of the expression is equal to *CE*.

**Domain ontology 3** *DisjointUnion(:Child ObjectComplementOf(ObjectComplementOf(:Boy)):Girl)*

## 4  Tool Features and Architecture

The prototype tool is implemented in Java language and consists of two parts which communicate through a socket.

The first part of the tool is a server, which is a runnable JAR file. The server performs operations on demand which are called by the client part. First of all, it calculates operations on the domain ontology and on the UML class diagram, such as:

(a) the normalization of the domain ontology,
(b) the normalization of the OWL representation of UML class diagram,
(c) the comparison of two sets of axioms (the normalized domain ontology and the normalized OWL representation of UML class diagram),
(d) the verification of the consistency of the modified domain ontology,
(e) detecting axioms that have caused inconsistency in the modified domain ontology,
(f) the calculation of the result of the diagram validation, and
(g) the calculation of suggested changes in the diagram based on the domain ontology (see Sect. 7).

The implementation of the server included two external libraries: the OWL API[2] and HermiT[3] OWL reasoner. The OWL API is a Java API for creating and modifying OWL 2 ontologies. HermiT reasoner is used to determine whether or not the modified OWL ontology is consistent in each iteration of the validation algorithm.

The second part of the tool is a client, which is a plug-into Visual Paradigm for UML. With the use of the plug-in the user can perform operations on demand from the server. The main feature of the plug-in is to conduct semantic validation of the drawn UML class diagram (an example is presented in Sect. 5). The additional features of the plug-in allow

(a) to display the normalized form of the input domain ontology (see Fig. 3),
(b) to display the OWL representation of the UML class diagram (see Fig. 4) and

---

[2]The OWL API—website: http://owlapi.sourceforge.net/.

[3]HermiT OWL Reasoner—website: http://www.hermit-reasoner.com/.

**Fig. 3** The example extract from the normalized input domain ontology

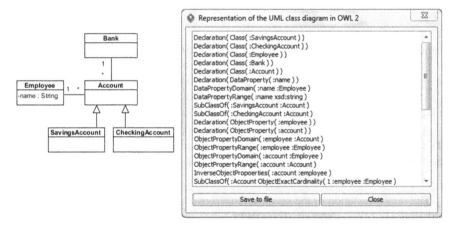

**Fig. 4** The example extract from the OWL representation of the presented UML class diagram

(c) to explicitly select elements from the domain ontology and place them on the class diagram as UML classes with attributes, associations, generalizations (between classes or associations), generalization sets and enumerations. The latest feature can be seen as a dictionary allowing to quickly present some parts of OWL domain ontology in the form of UML class diagram (see Figs. 5 and 6).

The user of the tool chooses the specific UML class diagram which needs to be validated and the specific domain ontology in OWL which will serve as a knowledge base. If the diagram is modified, the user may re-do the validation or conduct other calculations whenever needed.

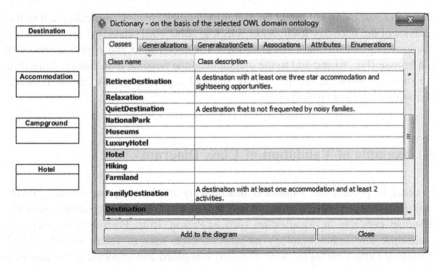

**Fig. 5** The classes tab of the dictionary based on the domain ontology

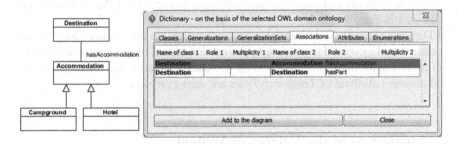

**Fig. 6** The associations tab of the dictionary based on the domain ontology

Figure 2 shows the initial form, which allows the user to select the input ontology and specify the diagram settings.

Figure 3 presents the result of normalization of the domain ontology selected by the user (here: travel ontology[4]). The normalized ontology is used as input to the method of validation but at any time it can also be viewed and saved to file for other needs.

Figure 4 shows an OWL 2 representation of the designed UML class diagram. In the method of validation the UML-OWL transformation is conducted in the background but at any time it can also be viewed and saved to file for other needs.

Figures 5 and 6 present a complementary feature of the tool—the possibility to automatically derive the needed part of UML class diagram directly from the selected domain ontology (here: travel ontology[4]). This feature is not used in the

---

[4]Travel ontology: http://protege.stanford.edu/junitOntologies/testset/travel.owl (accessed: 22.04. 2017).

method of validation but might be useful for a quick creation of the diagram or the analysis of the ontology itself. After selecting some UML classes (Fig. 5) from the ontology (e.g. *Destination*, *Accommodation*, *Campground* and *Hotel*), a modeller can browse (Fig. 6) and add to the diagram e.g. the possible associations defined in the ontology between the previously selected classes (e.g. association between *Destination* and *Accommodation*).

## 5 Example of Validation of Class Diagram

In order to present the validation functionality of the tool, two existing OWL ntologies were selected: Organizations[5] and Teams.[6] The Organizations ontology describes concepts in organization such as Person, Role, Position and Organizational Unit. The Teams ontology includes concepts such as Team, TeamType and TeamRoleType. Ontologies were manually combined and imported to the tool as a domain ontology.

Figure 7 presents the example UML class diagram which needs to be validated against the selected combined domain ontology. The results of the validation are illustrated in Figs. 8 and 9.

Diagram elements (transformation axioms) which are not contradictory to the domain ontology should be verified by the domain expert, because the axioms were not defined in the domain ontology. In the presented diagram *CrossFunctionalTeam* and *duration* attribute of *TemporaryTeam* are such examples.

## 6 Limitations of the Validation Method

The tool and the proposed method of semantic validation of UML class diagrams have some limitations:

- The method is limited to validate only static aspects of UML class diagrams, and the behavioural features, such as class operations, are omitted. This limitation is motivated by the fact that the OWL 2 ontologies contain classes, properties, individuals, data values, etc. but does not allow to define any operations that may be directly invoked e.g. on the individuals.
- Some elements of UML class diagrams, e.g. n-ary associations, compositions, etc. are not fully translatable into OWL 2 (e.g. properties in OWL 2 are only

---

[5]Organization OWL ontology: https://raw.githubusercontent.com/isa-group/cristal/master/ral-ontology/organization.owl (accessed: 22.04.2017).

[6]Teams OWL ontology: https://raw.githubusercontent.com/isa-group/cristal/master/ral-ontology/teams.owl (accessed: 22.04.2017).

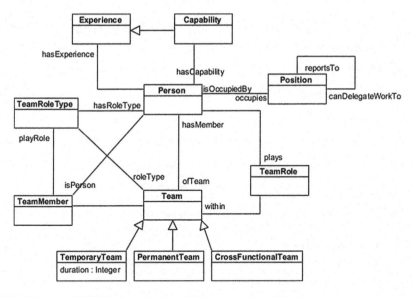

**Fig. 7** The example UML class diagram which needs to be validated

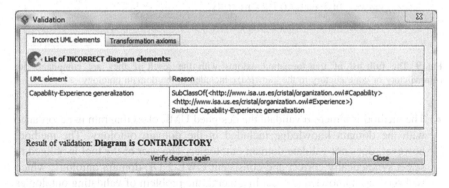

**Fig. 8** The detected contradictory axiom

binary relations). The elements can only be partly translated (e.g. n-ary association can be partly translated with the use of the pattern proposed in [21]). However, the partial translation is still justified for the purpose of diagram validation. This limitation is caused by the fact that UML and OWL standards differ from each other, e.g. [6, 7].

- There is a limitation, which requires all class attributes in one UML class diagram to be uniquely named. This limitation is caused by the fact that different UML attributes of the same name would be mapped to one OWL property, which should change the semantics.

**Fig. 9** The full list of transformation axioms with the result if they are compliant, not contradictory or contradictory to the knowledge included in the domain ontology

- The method is aimed to validate the designed UML class diagram in accordance with the domain knowledge included in the domain ontology. The method, however, does not validate domain ontologies. The user of the tool is obliged to select syntactically correct and consistent OWL 2 DL domain ontology, which will serve as a knowledge base. In general, the problem of validating ontologies requires a comparison of the ontologies with an expert knowledge either provided by domain experts or included in another source of domain knowledge.

## 7 Ontology-Based Suggestions for Diagram Corrections

Having UML class diagram compliant (or at least not contradictory) with the domain ontology does not assure that the diagram most adequately reflects the selected aspect of the domain. The proposed suggestion of changes are aimed to be reported to the modeller in order for him or her to decide whether the model should incorporate the specific correction. Depending on the context, sometimes it might

**Table 1** Missing attributes of UML class

| UML class diagram element | Domain ontology | Suggested correction |
|---|---|---|
| A<br>b : B | A<br>b : B<br>c : C | Insert missing attribute to the UML class |

be useful not to present unnecessary details in the diagram. Missing classes and associations between the classes are purposely omitted from the suggested changes, due to the fact that it seems to unnecessarily extend the diagram.

For a clarity of reading, the following diagram elements (identified to be corrected) are displayed using the same notation—UML (the elements could be presented in OWL as well) (Tables 1, 2, 3, 4, 5, 6 and 7).

**Table 2** Lack of specified multiplicity of attributes or not precise multiplicity

| UML class diagram element | Domain ontology | Suggested correction |
|---|---|---|
| A<br>b : B | A<br>b : B [N..M] | Insert missing multiplicity of attribute |
| A<br>b : B [K..L]<br><br>$K \geq N$ and $L \leq M$ and $K+L \neq N+M$ | A<br>b : B [N..M] | Precisely specify multiplicity of attribute |

**Table 3** Lack of specified multiplicity of association ends or not precise multiplicity

| UML class diagram element | Domain ontology | Suggested correction |
|---|---|---|
| A ———— B | A ———— B<br>N..M | Insert missing multiplicity of association |
| A ———— B<br>K..L<br><br>$K \geq N$ and $L \leq M$ and $K+L \neq N+M$ | A ———— B<br>N..M | Precisely specify multiplicity of association |

**Table 4**  Not complete enumeration

| UML class diagram element | Domain ontology | Suggested correction |
|---|---|---|
| <<enumeration>> A, a, b | <<enumeration>> A, a, b, c | Insert missing enumeration literals |

**Table 5**  Missing generalization relationship between the classes on the diagram

| UML class diagram element | Domain ontology | Suggested correction |
|---|---|---|
| A — B | A ◁— B | Insert missing generalization between the classes |

**Table 6**  Not precise constraint of incomplete generalization set

| UML class diagram element | Domain ontology | Suggested correction |
|---|---|---|
| A, {incomplete}, B, C | A, {complete}, B, C | Insert complete constraint to generalization set |

**Table 7**  Not precise constraint of overlapping generalization set

| UML class diagram element | Domain ontology | Suggested correction |
|---|---|---|
| A, {overlapping}, B, C | A, {disjoint}, B, C | Insert disjoint constraint to generalization set |

# 8 Conclusions

The article presents the current features of the tool for semantic validation of UML class diagrams. A knowledge base for the validation of the diagram is the domain ontology expressed in OWL 2 selected by the user. Reusing domain ontologies opened the opportunity to reduce the necessity of a constant involvement of domain experts in the process of diagram validation. With the use of the proposed method, the expert involvement is targeted towards unifying the vocabulary between the class diagram and the ontology, and not towards the validation process itself.

The method of validation outlined in [5] was extended with some additional steps, among which the normalization of the ontologies is the most important. It allows to easily compare the two OWL ontologies which were originally built of syntactically different axioms. The current research is focused on defining a complete list of normalization rules for OWL 2 DL ontologies and to incorporate them in the prototype tool. The tool will also be extended with further functionalities focused on offering better user experience in interpreting the results of validation. For example, an intended function is a colour distinction of the elements which are semantically incorrect or are suggested for the modification based on the knowledge included in the ontology. Additionally, ontology-based suggestions of diagram corrections are planned to be further extended and detected for the element(s) selected by the user on the diagram.

# References

1. OWL 2 Web Ontology Language Document Overview (Second Edition). W3C Recommendation 11 December 2012. https://www.w3.org/TR/owl2-overview/ (2012)
2. OWL 2 Web Ontology Language. Structural Specification and Functional-Style Syntax (Second Edition). W3C Recommendation 11 December 2012. http://www.w3.org/TR/owl2-syntax/ (2012)
3. Horrocks, I., Kutz, O., Sattler, U.: The even more irresistible SROIQ. In: Proceedings of the 10th Int. Conf. on Principles of Knowledge Representation and Reasoning (KR 2006). AAAI Press, pp. 57–67 (2006)
4. OWL 2 Web Ontology Language New Features and Rationale (Second Edition) W3C Recommendation 11 December 2012. https://www.w3.org/TR/owl2-new-features/ (2012)
5. Sadowska, M. and Huzar, Z.: Semantic validation of UML class diagrams with the use of domain ontologies expressed in OWL 2. Software Engineering: Challenges and Solutions. Springer International Publishing, pp. 47–59 (2017)
6. OMG: Unified Modeling Language, Version 2.5, Doc. No.: ptc/2013-09-05. http://www.omg.org/spec/UML/2.5 (2015)
7. SPARQL 1.1 Overview: W3C Recommendation 21 March 2013. https://www.w3.org/TR/sparql11-overview/ (2013)
8. Xu, Z., Ni, Y., He, W., Lin, L., Yan, Q.: Automatic extraction of OWL ontologies from UML class diagrams: a semantics-preserving approach. World Wide Web 15(5-6), 517–545 (2012)
9. El Hajjamy, O., Alaoui, K., Alaoui, L., Bahaj, M.: Mapping UML To OWL2 Ontology. J. Theor. Appl. Inf. Technol. 90(1), 126 (2016)

10. Zedlitz, J., Jörke, J. & Luttenberger, N. From UML to OWL 2. Knowledge Technology, pp. 154–163. Springer, Heidelberg (2012)
11. Khan, A.H., Porres, I.: Consistency of UML class, object and statechart diagrams using ontology reasoners. J. Vis. Lang. Comput. **26**, 42–65 (2015)
12. Khan, A.H., Rauf, I., Porres, I.: Consistency of UML Class and Statechart Diagrams with State Invariants. Modelsward, pp. 14–24 (2013)
13. Zedlitz, J., Luttenberger, N.: Transforming between UML conceptual models and OWL 2 Ontologies. In: Terra Cognita 2012 Workshop, vol. 6 (2012)
14. Bahaj, M., Bakkas, J.: Automatic conversion method of class diagrams to ontologies maintaining their semantic features. Int. J. Soft. Comput. Eng. (IJSCE) **2** (2013)
15. OWL 2 Web Ontology Language Profiles (Second Edition). W3C Recommendation 11 December 2012. https://www.w3.org/TR/owl2-profiles/ (2012)
16. OWL2UML, http://apps.lumii.lv/owl2uml/ (2009)
17. ProtégéVOWL: VOWL Plugin for Protégé. http://vowl.visualdataweb.org/protegevowl.html (2014)
18. VOWL: Visual Notation for OWL Ontologies. Specification of Version 2.0. http://vowl.visualdataweb.org/v2/ (2014)
19. Bārzdiņš, J., Bārzdiņš, G., Čerāns, K., Liepiņš, R., Sproģis, A.: OWLGrEd: a UML style graphical editor for OWL. In: Proceedings of ORES-2010, CEUR Workshop Proceedings, vol. 596 (2010)
20. Bogumiła, H.: Towards automatic SUMO to UML translation. From Requirements to Software, Research and Practice, pp. 87–100 (2015)
21. Noy, N., Rector, A.: Defining N-ary Relations on the Semantic Web, W3C Working Group Note 12 April 2006. http://www.w3.org/TR/swbp-n-aryRelations/ (2006)

# Ensuring the Strong Exception Safety

Piotr Kosiuczenko

**Abstract** Algorithms for the transaction rollback belong to the fundamentals of database theory and practice. Much less attention has been paid to the method rollback in the realm of object-oriented systems. This issue was studied in the context of exception handling. In a previous paper, a new algorithm for computing old attribute values has been proposed. In this paper that algorithm is used to implement a procedure for method rollback. It is shown that this procedure ensures the so called strong exception safety.

## 1 Introduction

Entire elimination of software bugs in complex software systems is impossible. Thus, it is necessary to have mechanisms for the system recovery from errors. There are various methods for assuring a correct operation. Exceptions handling is the most fundamental mechanism for dealing with situations which pervert the normal flow of program execution or contradict system consistency requirements. Exceptions are used to signal that something went wrong and that a routine could not execute normally. The associated handler mechanism is used then to manage the situation.

In case of databases, there are two basic approaches for handling transaction failures (cf. e.g. [11]). The deferred update technique postpones updating the real database until a transaction reaches the commit point. All transaction updates are recorded in a transaction workspace; thus, undoing changes is unnecessary in case of a transaction failure. The immediate update approach allows real database updates before a transaction commits. Those updates are usually recorded in a log file before they are applied to the database. If a transaction fails, then the performed changes are rolled back according to the log. Some techniques combine those two approaches. In SQL, the rollback is a command that causes all data changes since the start of a transaction to be discarded. In most SQL dialects, this command is connection specific,

P. Kosiuczenko (✉)
Institute of Information Systems, WAT, Warsaw, Poland
e-mail: piotr.kosiuczenko@wat.edu.pl

© Springer International Publishing AG 2018
P. Kosiuczenko and L. Madeyski (eds.), *Towards a Synergistic Combination of Research and Practice in Software Engineering*, Studies in Computational Intelligence 733, DOI 10.1007/978-3-319-65208-5_5

i.e. rollback made in one connection will not affect any other connections. A cascading rollback occurs in database systems when a rollback of one transaction causes rollbacks of dependent transactions. In Java method `rollback` allows the transactions rollback via interfaces `Connection` and `UserTransaction`. It reverts all changes made since the last transaction was committed or rolled back and releases all database locks currently held by this connection or thread respectively.

In the realm of object-oriented systems, the idea of method rollback is considered in the context of exception handling (see [20]). Exceptions can be thrown at any point of method execution. In this case it is either handled or the call-stack is unwound. The exception may be thrown when the system state is inconsistent. In the literature, some consistency criteria in the presence of exceptions have been defined (cf. [1, 20]). Those criteria range from no throw guarantee, via the so called strong exception safety, to the requirement that no resources are leaked. It is not always possible to guarantee that no exception will be thrown. The second criterion requires that in case of exception a method must be rolled back; this implies performance and memory problems. Only the weakest criterion is considered to be reasonably implementable in all cases [1]. It should be pointed out that a cascading transactions rollback is a different problem from method rollback. Simultaneous execution of multiple transactions happens as if they were performed in a sequence, i.e. as if one was started after the completion of another. In case of nested method calls, method called during execution of another one must be completed before the calling method returns.

The problem of reverting transactions is closely related to the problem of old value archiving. In the seminal paper [8], the notions of partially persistent and persistent data structures have been defined. The first one corresponds to structures for which every consecutive version can be accessed and the last version can be modified. The second one corresponds to structures for which in addition all versions can be modified. In both cases, the access time to old values is at least logarithmic. This approach does not deal with method calls, which per se can be nested. In the paper [7], the notion of semi persistent data structures has been coined. Those structures ensure that all forms prior to unterminated methods/procedures can be accessed and modified; however, the authors have not explained how this can be achieved. The notion of sufficiently persistent structures has been introduced [15]. It requires that all states of modified structures prior to not terminated methods can be accessed, but only the most recent one modified. It was shown that the cumulative access time to attribute values from before the last method execution is constant. In that paper, the correctness of the old value retrieval algorithm has been proved.

In the realm of object-orientation the so-called design by contract is used for a system specification [16]. There exist various languages for contractual specification. We mention here only OCL [4, 19], Eiffel [17], JML [10] and Spec# [6]. Data integrity and system consistency is defined in terms of conditions called invariants. A method pre-condition defines situations when the method can be called. A method post-condition specifies its results. Its evaluation requires the computation of attribute values from before the method execution. Those values are returned by operators `@pre` of OCL and old of Spec# and JML. In these languages, the computation of old attribute values is based on the idea of saving those values in the pre-state. However not all such values can be pre-computed. As a result, this approach supports only a restricted form of contracts (cf. [13]). In general, deep cloning is computationally expensive, especially in the case of collection types. Such clones are the major slowdown factor in a contract evaluation (cf. [9]). They pose also logical problems, since reference identity cannot be used for object comparison.

In this paper, we propose a method ensuring the so called strong exception safety. This method is based on our previous result concerning the computation of old attribute values [14]. The algorithm proposed in that paper overcomes above mentioned problems. It allows to access @pre-values during post-condition evaluation as needed. Its execution does not increase the complexity class of post-condition evaluation as if the computation of @pre-values had a constant time. Since old attribute values are not recorded in advance, there is neither need to clone system states nor to restrict the post-condition syntax. In this paper, we use this algorithm to revert the modified attribute values. However, it cannot be applied directly, since we need to know which objects are modified during a method call. Since method calls can be nested and rolling back one method requires rolling back all methods which were called during the failed method execution. For every attribute, we define a history-stack whose elements are snapshots consisting of an old attribute's values and its time-stamps being the calls number. Basically, if an attribute is set for the first time during a method execution and the stack does not include timely snapshots, its previous value is pushed on the stack with a time-stamp being the number of the currently executing method. We define also clean operations which remove outdated snapshots from history-stacks. Objects modified during a method execution are registered in a special set. After the method returns, the union of this set and of the modified object set corresponding to the calling method is formed.

Proposed algorithm rolls methods which have thrown exceptions back to the point where they were called, and in this way, ensures the strong exception safety. This algorithm does not increase the complexity class of instrumented methods. We present also test results concerning the time- and space-overhead caused by our method. It turns out, that the overhead is around 5.

This paper is organized as follows. In Sect. 2, we discuss different approaches to and levels of exception safety. In Sect. 3, we present an example illustrating our approach. In Sect. 4, we present an aspect-oriented implementation of the rollback algorithm. In Sect. 5, we discuss the time overhead of our method. Section 6 concludes this paper.

## 2 Exception Safety

Programming languages differ in their support for exception handling. Most languages provide exception handling mechanisms. When an exception is thrown, the call-stack is unwound until a proper exception handler is found or the processing is aborted and an error message displayed. An exception is handled by starting a specific subroutine. Depending on the situation, the handler routine may later resume the execution at the original location using saved information.

During a method execution, files may be opened and not closed prior to an exception. Similarly, a method can acquire an internet connections. It is important to ensure that after throwing an exception, there are no resource leaks and thus allocated resources, such as files, locks, network connections and threads, are eventually released. An operation is said to be exception-safe if runtime failures during its execution does not produce undesired effects, such as inconsistent memory states or invalid output. Various exception safety levels have been defined in the literature (cf. [1, 20]).

1. The minimal exception safety condition requires that failed method execution does not cause a crash and that no resources are leaked
2. The basic exception safety condition requires preservation of invariants and prohibits resource leaks
3. The strong exception safety guarantee ensures that a failed operation has no side effects
4. The no throw guarantee ensures that operations are guaranteed to succeed and all acquired resources released

The minimal guarantee ensures that every resource that was acquired is eventually released, i.e. files are closed, locks are freed, network connections are closed, etc. It is a minimum standard for exception-safety. The basic condition implies the preservation of invariants and prohibits memory leaks, but allows side effects. Thus, system states are guaranteed to be consistent at the end even if they are different from before the partial execution. However, this guarantee is not very useful, since it is not clear what state the component is in after an exception is thrown.

The strong exception safety guarantee is called sometimes no-change guarantee or rollback semantics. It implies that there are no side effects. It is equivalent to the so-called commit-or-rollback condition. This condition requires that if an exception is thrown during a method execution, then it does not cause resource leak and the states of all objects that the method modified are restored to the state they had before the method was called (see [3]). An operation that has no effects if failing has the clear benefit that the system state is known and it can be used as before the operation's execution [1].

The no throw guarantee is called sometimes failure transparency condition. It implies that no exception is throw by the operation even in presence of exceptional situations. It would be ideal for a client. However, in general it is not possible to achieve. The strong guarantee is a possible alternative. Abrahams [1] points out that in case of collections its achievement may have a high cost. First the collection needs to be copied, then the operation is executed on the copy and finally the modified copy replaces the original collection. It is a time- and space-consuming procedure. An insertion near the end of a vector, which may initially cause only a few copies, requires every element to be copied. Performance requirements are the reason why the basic guarantee is considered to be the 'natural' level of safety for this operation [1]. Some authors pointed out that in the present state of data structures it is impossible to implement a method to deliver either the basic or strong exception safety guarantees if the behaviour in the presence of exceptions is not known (see [1, 12]).

There is a pattern ensuring the strong exception safety in the case when the modified objects can be figured out prior to a method execution [12]. It is based on the pattern called 'Execute Around Method' introduced by Beck [5]. The idea is that methods that can throw exceptions are executed first to obtain results and the results are stored. Then all resources allocated in the preceding step get released if the previous block was left due to an exception. Finally, the temporary results are assigned to the actual attributes using exclusively operations that cannot throw exceptions. In general, the pattern called RAII (Resource Acquisition Is Initialization) for resource deallocation [20] and the try-catch-finally statement form the backbone of exception safe programming. Some research has also been done on a static code analysis for the strong exception safety guarantee. In the papers [2, 18] an algorithm was proposed which checks if C++ code fulfils this guarantee.

The weak exception safety guarantee seems to be easier to implement than the strong one. In most applications, the creation of text files and the opening of connections is done sparingly. Thus, in practice, weak exception safety is easier to achieve. However, from the formal point of view, its implementation suffers the same problems as the implementation of the strong requirement. A method can for example create text files as it creates objects. For every method, one can simulate an object creation and modification with creation of new text files and their wrapper objects. In this case figuring out which text files were created and opened during a method execution is equally difficult as figuring out which objects were modified.

## 3 Example

In this section, we present a list example which illustrates our approach. The list is composed of an anchor object of class List and some elements instantiating class Element. List objects have attribute first storing the first element of the list. There is also method insertEl(Element el) for inserting elements into the list. Element objects have attribute val storing integers and attribute next pointing to the next element.

We implement insertEl in a very inefficient way. An element el is inserted into a list if the list is empty or el.val is smaller than the value of its first element. If not, then list elements are removed if the value of the first element is larger than the value of the inserted element, then el is inserted and afterwards all removed elements are inserted back.

To ensure throughout testing, we implement the method so that at any point an exception can be thrown with some nonzero probability. The insertion of new elements in a list reminds a Sisyphus work, since at any point exception can be thrown and the call-stack unwound arbitrarily deep before the insertion is retried.

All elements inserted during a failed method execution are removed, and afterwards they need to be reinserted. Unlike the mythological Sisyphus, the element insertion will be completed with probability one, even though this can take arbitrarily long. Every method call is placed within a try-clause and accompanied by a corresponding catch clause. If an exception is caught, then with some probability it is re-thrown or it is handled by restarting the method with the same argument. For this method to operate correctly, changes made by a failed insert must be discarded after the exception is thrown.

Below we present a Java implementation of classes `List` and `Element`. It should be noted that during execution of `insertEl` several other calls of this method can be made if the inserted element is not smaller than the first element saved in the list. The call-stack can contain several calls to this method. On the other hand during execution of this method, some calls may be terminated. This implies that a proper stamping policy is needed to guarantee that the method calls are numbered uniquely and that one needs a proper logic to figure out which snapshots are valid.

```
public class List {                        } else {
  private Element first;                     setFirstWithExc(getFirst()
  static Random random =                                 .getNext());
              new Random();                   try { insertEl(el);
  public void                              } catch(MyException ex){
        insertEl(Element el) {               insertEl(el);
    if(getFirst() == null) {                 }
      setFirstWithExc(el);                   try{insertEl(oldFirst);
      el.setNextWithExc(null);             } catch(MyException ex){
    }                                          insertEl(oldFirst);
    else if(el.getVal() <=                   }
        getFirst().getVal()) {             }
    el.setNextWithExc(getFirst()); }
      setFirstWithExc(el);
```

Method `setFirstWithExc` wraps `setFirst` so that before and after its execution an exception is thrown with some probability. The same holds for method `setNextWithExc` in the class `Element`. Thus, at any point of `insertEl` execution an exception can be thrown. If the changes done by the failed operation were not undone, then `insertEl` would produce incorrect results due to for example elements removal. Method `setFirstBis` is meant to be used within aspects. Class `Element` defines list elements and method `setNextWithExc`.

```
public class Element {                            (Element next){
  private int val = 0;                       List.rundomizeEx(next,0.03);
  private Element next = null;               this.setNext(next);
  public void setNext                        List.rundomizeEx(next,0.03);
            (Element next)                  }
  { this.next = next;                        //...
  }                                       }
  public void setNextWithExc
```

It should be pointed out that monitored set operations do not throw exceptions. As mentioned above, `setFirstBis` and `setNextBis` are defined to be used in aspects to avoid interference of the aspect routines with aspects themselves.

## 4 Implementation of the Rollback Algorithm

In this section, we present an AspectJ implementation of the rollback algorithm and show how to use it to make methods strong exception safe. In our approach, we avoid the problems of deep cloning. However, we need a method to archive objects and reassemble them from their histories.

### 4.1 Implementation Idea

This approach is based on the algorithm for computing old attribute values proposed in [14]. That algorithm superimposes the so-called fat structure [8] on classes, but handles attribute modification in a way different from the persistent data structures. The modification of attributes should be treated in a special way due to method returns. For the method rollback, we need a method for a proper registration of modified objects. We need also a method for restoring old attribute values.

We superimpose history attributes and methods for restoring values of modified attributes. We define so called pointcuts for registering relevant events, and advices, for handling those events. We use a kind of time-stamp to be able to reassemble objects states which existed at different times, but unlike the numbering used in [8] those stamps refer to method calls. Similarly, the numbering used here is not monotone due to method returns and therefore it needs a special treatment. We call a method which has to be made strong exception safe 'relevant'. Old values of attributes modified during a failed execution of a relevant method must be restored. Thus, we need to archive values of attributes which can be potentially modified during a relevant method call. Such attributes can be identified using a static code analysis. A method is made exception safe by superimposing the above-mentioned structure. If the method throws an exception, then attributes modified during its execution are restored. It is possible, since we register modified objects in a special set. For those objects the restore methods are executed.

The class instrumentation looks as follows. If in class C an attribute a of type T can be modified by a relevant method, then we add set-method `setA`, if not already present, and `setABis`. We superimpose on class C a history attribute `aHIST` of type `Stack<SnapshotVal<T>>` to store the history of a. We superimpose also pointcut which listens to changes of a and an advice which is responsible for saving its old values. Values of attribute a are restored by the superimposed method `restoreA()`. The superimposition is done by aspect `ArchiveC`.

Parametric class `Stack<S extends Snapshot>` implements stacks and contains method `clean()` for the removal of outdated snapshots from attribute histories. Objects of parametric type `Stack<SnapshotVal<T>>` store histories of attributes in the form of a stack. Parametric class `SnapshotVal<T>` defines attribute snapshots, i.e. a temporary value of an attribute plus the corresponding time-stamp, or as we call it later meter-reading. As mentioned above, this time-stamp correspond to the call number of the method which modified the attribute.

Class `Meter` stores information about the number of relevant method calls on the program stack and the lists of objects modified during execution of those methods. `MeterAspect` handles calls to relevant methods. Class `Archive` includes the core logic for value archiving and implements the generic method `restored Val()` used in the implementation of methods such as `restoreA()`.

We explain our algorithm using the list example. We make `insertEl` strong exception safe. This method modifies attributes `next first` and `insertEl`. When it fails, their values must be restored. Thus, they must be archived. Attribute `val` is not modified and consequently does not need to be archived.

## 4.2   Implementation

In this subsection, we present the implementation. We present the above-mentioned classes and explain how the superimposition works.

Class `Meter` handles time-stamps. They correspond to the number of calls of `insertEl`. The number of the last call is stored in attribute `meterValue`. `Meter` handles also lists of objects modified during calls of this method. Those lists are stored as attributes of objects of class `SnapshotModif`. These objects are stored in stack `meter`. The goal is to register objects modified during a method call.

```
public abstract class Meter {
  static int meterValue = 0;
  static final
    Stack<SnapshotModif> meter
    =new Stack<SnapshotModif>();
  public static int getReading(){
  if(meter.size() == 0) return 0;
  return meter.top().meterReading;
  }
  public static void
          increaseMeter(){
    ++meterValue;
    SnapshotModif n =
              new SnapshotModif
                        (meterValue);
      meter.push(n);
  }
  public static void
            decreaseMeter() {
    meter.pop();
  }
  public SnapshotModif
            getTopSnapshot() {
    return meter.top();
  }
}
```

`MeterAspect` is a generic aspect implementing the logic concerning method rollback. When a monitored method is called, in this case `insertEl`, the meter is increased, a new object of class `SnapshotModif` storing lists of modified objects is created, stamped with the current meter-reading and pushed onto the meter-stack by executing `increaseMeter()`. Modified objects are registered in the corresponding lists `modifiedFirst` and `modifiedNext`, respectively. After `insertEl` returns, if the stack is not empty, then the lists of modified objects are moved down the stack. This is necessary, since if the method fails at a later point in time we need to know which objects were modified. It should be noted that we do not clean attribute histories here to avoid square number of object cleans. If the stack is empty (meaning that `main` is executing), then all history attributes are emptied and `meterValue` is set to 0, since there is no active call of a relevant method on the program stack. Then the lists of modified objects are emptied. In case of exception, if the meter-stack is not empty, then attributes modified during the failed method execution must be set back. After restoring old attribute values, modified object lists are not pushed down. The case of meter-stack containing only method `main` is handled in a similar way to method return.

```
public aspect MeterAspect {                     Meter.meterValue = 0;
  declare precedence: *,                      }
                 MeterAspect;               }
  public pointcut ins1() :                  after() throwing : ins1(){
    execution(void insertEl                   SnapshotModif sm =
                 (Element));                       Meter.meter.top();
    before() : ins1() {                       if(Meter.meter.size()>0)
      Meter.increaseMeter();                  { for(List o :
    }                                              sm.modifiedFirst)
    after() returning : ins1() {              o.restoreFirst();
      SnapshotModif sm =                      for(Element o :
              Meter.meter.top();                  sm.modifiedNext)
      Meter.decreaseMeter();                  o.restoreNext();
      if(Meter.meter.size() > 0)             }
      /*after ret., push                      else //meter.size()==0
       modif. objects' lists down*/{         { for(List an :
      Meter.meter.top().modifiedFirst              sm.modifiedFirst)
          .addAll(sm.modifiedFirst);          an.firstHIST.empty();
      Meter.meter.top().modifiedNext           for(Element el :
          .addAll(sm.modifiedNext);               sm.modifiedNext)
      }                                        el.nextHIST.empty();
      else /*Meter.meter.size()==0*/{        sm.modifiedFirst.
      for(List an : sm.modifiedFirst)          removeAllElements();
              an.firstHIST.empty();         sm.modifiedNext.
      for(Element e :sm.modifiedNext)          removeAllElements();
              e.nextHIST.empty();           Meter.meterValue = 0;
      sm.modifiedFirst.                      }
              removeAllElements();           Meter.decreaseMeter();
      sm.modifiedNext.                   }
              removeAllElements(); }
```

`Archive` is a generic class implementing the core logic for handling old attribute values and value restoring. `restoredVal` returns the old value of an attribute and modifies its history appropriately. It contains two parameters: `val` corresponding to the current value of an underlying attribute and `st` corresponding to the attribute's history-stack. If `st` is empty, then the attribute was not modified and `val` is returned. If not, then outdated snapshots are removed from the attribute's history-stack. If the topmost element in the stack has a stamp smaller than the current meter-reading, then it means that during the recent method call the value was not modified and val is returned. If not, then the attribute was modified during this or a later, but already terminated, method call. In this case, the topmost value is returned and the history stack is popped. Method getLastUpdateTime returns last update time of an attribute. This value is taken from the topmost snapshot in the attribute's history-stack if it is not empty; in the other case 0 is returned.

doArchiving is meant for storing attribute's snapshots on an underlying attribute history stack and registering the modified objects. This method has four parameters: st corresponds to the history stack of the underlying attribute, modTop corresponds to the topmost list of objects for which the underlying attribute was modified, modBottom corresponds to the bottommost list, target corresponds to the object for which the relevant attribute is above being modified, cur corresponds the current attribute's value. If stack st is empty, then a new attribute snapshot is created with the current meter-reading, put onto the stack and the object is registered in the top and bottom lists of modified objects. The object is saved at the bottom, since it is its modification for the first time. If st is not empty, then it is checked if the attribute's value was saved during the current or an already terminated method execution. If it not saved, then the target object is registered in modTop. If the value is not on st, then a new attribute snapshot with the current meter-reading is created and pushed on the stack. If it is saved, then only the meter-reading in the corresponding snapshot is updated.

```
public class Archive {
   static <T> T restoredVal(
      Stack<SnapshotVal<T>> st,
   T val) {
      T t = val;
      st.clean();
      if(st.size() == 0 ||
         st.topReading() <
         Meter.getReading())
      return t;
      else t = st.top().value;
      if(st.size() > 1 &&
         st.topReading() >=
         Meter.getReading())
      st.pop();
      return t;
   }
   static <T> int
   getLastUpdateTime(
      Stack<SnapshotVal<T>> st){
      if(st.size() ==0) return 0;
      st.clean();
   return st.top().meterReading;
   }
   static <T, S> void
      doArchiving(
          Stack<SnapshotVal<S>> st,
          Vector<T> modTop,
            Vector<T> modBottom,
          T target, S cur) {
      if(st.size() == 0)
         //history is empty
         { st.push(new
            SnapshotVal<S>(cur,
            Meter.getReading()));
            modBottom.add(target);
            modTop.add(target);
      }
      else //not empty
      if(Meter.getReading()
         >st.top().meterReading){
         modTop.add(target);
         //register modif. object
         if(st.top().value != cur)
           st.push(new
             SnapshotVal<S>(cur,
             Meter.getReading()));
      else st.top().meterReading
         = Meter.getReading();
      }
   }
}
```

It should be noted that we need to update the meter-reading of the actual snapshot if the archived value coincides with the attribute value before it is set for the first time during a method execution. It is because the attribute can be changed several times during the method execution. Similarly, we need to register the object modification.

Abstract class `Snapshot` has attribute `meterReading` for saving the current meter-reading of a snapshot, i.e. the number of the call being executed at its creation time. `SnapshotModif` extends `Snapshot` and is meant for storing the lists of objects for which attributes requiring archiving were modified during a method execution. In this case, these are attributes `first` and `next`. Inherited attribute `meterReading` stores the current meter-reading. `SnapshotVal` is a parametric class extending `Snapshot`; its objects are used to store attributes' snapshots. Attribute `value` stores the attribute's value.

```
public abstract class Snapshot{         }
   int meterReading;                   }
}                                       public class SnapshotVal<T>
public class SnapshotModif                      extends Snapshot{
         extends Snapshot{               T value;
   public Vector<Anchor>                 public SnapshotVal(T el,
   modifiedFirst =                                    int t){
      new Vector<Anchor>();               value = el;
   public Vector<Element>                 meterReading = t;
   modifiedNext                          }
      = new Vector<Element>();           //... other constructors
   public SnapshotModif(int i){         }
   meterReading = i;
```

Parametric class `Stack` implements a stack with methods for popping and pushing snapshots. It is implemented here using `Vector`, but other collections from the Java API standard library can be used as well. Apart of above mentioned methods, it contains method `subTop` for returning the element below the topmost position. Method `clean()` is used for removing outdated snapshots; i.e. snapshots located at the top of the stack whose meter-reading is larger than the current meter reading and who have a direct follower, returned by `subTop()`, with meter-reading larger than or equal to the current meter reading. It should be noted that the cleaning must be done in a while loop, since in general it is not enough to remove only the topmost outdated snapshot. Methods `top()`, `pop()`, `size()` are skipped here due to their obvious implementation. We have skipped also method `bottom()` that returns the bottommost element and `discharge()` that removes all elements from the stack.

```
public class
   Stack<S extends Snapshot> {
 Vector<S> stack =
              new Vector<S>();
 public int topReading() {
   if(stack.isEmpty()) return 0;
   return top().meterReading;
 }

     stack.remove(
          stack.size()-1);}
 //..
}

public aspect ArchiveList {
 public
   Stack<SnapshotVal<Element>>
     List.firstHIST = new
 Stack<SnapshotVal<Element>>();
 int List.getLastUpdateTime(){
   return
   Archive.getLastUpdateTime(
              firstHIST);
 }
 void List.restoreFirst() {
   setFirstBis(
     Archive.restoredVal(
     firstHIST, getFirst())));
 }
```

```
public S subTop() {
  return
  stack.get(stack.size()-2);
}
public void clean() {
  while(size() >= 2 &&
    subTop().meterReading
      >= Meter.getReading())

pointcut modFirst
              (List target) :
   target(target) &&
     call(void
     List.setFirst(Element));
before(List target) :
         modFirst(target) {
   if(Meter.getReading() > 0)
   {  Archive.doArchiving(
      target.firstHIST,
      Meter.meter.top().
              modifiedFirst,
      Meter.meter.bottom().
              modifiedFirst,
      target,
      target.getFirst());
   }
 }
}
```

For every class C with attributes requiring archiving, we introduce a dedicated aspect ArchiveC which superimposes corresponding the history attributes and the methods returning old values of attributes. This aspect also manages setting of attributes. In the example, for the classes Anchor and Element we introduce aspects ArchiveAnchor and ArchiveElement. first is the only attribute of class Anchor which must be archived. For this attribute method restore First() is superimposed on Anchor. It returns the old value of first and cleans its history stack. This method is implemented with the help of restored Val, getLastUpdateTime and setFirstBis. We cannot use here set First, since it would cause undesired interference within the aspect. Every manipulation of first by setFirst is detected by pointcut modFirst. If the current meter-reading is larger than 0, meaning that there is a relevant method on the stack, then the archiving is performed by doArchiving. Above we presented the aspect ArchiveAnchor. Aspect ArchiveElement, corresponding to class Element, can be implemented analogously.

# 5   Time Overhead

In this section, we discuss the issue of time- and space-overhead due to the application of our method. Direct comparison is possible between the execution of a code in the un-instrumented and the instrumented form. In the paper [15], it has been shown that the attribute archiving does not increase the complexity class of instrumented methods and that the cumulative access time to attribute values from before a method execution is constant, in contrary to partially persistent structures, where in general the access time is logarithmic. Therefore, the time cost of a method rollback is at most linear in respect to the number of objects modified during a failed method execution. However, in some cases, where the execution is based mostly on numeric computations or on a manipulation of a few attributes, the rollback can be logarithmically related to the execution time. This is for example the case of the recursive algorithm handling Towers of Hanoi. It can be shown that the rollback time needed in case of a failed move ring operation is logarithmically related to the algorithm execution time. This is due to the fact that the time complexity of this algorithm is exponential in the number of rings, a list with n rings can be implemented using $n + 3$ objects, the recursion depth is at any point at most n and every call modifies only those objects.

We have measured the execution time and memory use of an un-instrumented program and the overheads incurred by the instrumentation. In particular, we have measured the number of objects being created, their size and time spend in certain code. The performance of `insertEl` has been tested for lists of various lengths. We have used Eclipse Test and Performance Tools Platform (TPTP), an Eclipse plugin that for performance-testing and profiling of Java applications [21]. The tests have shown that the slowdown factor is around 7.5 if AspectJ is used to instrument the code. The memory consumption has been increased 5 times.

It should be noted that attribute modifications as such are computationally inexpensive. However, their monitoring with pointcuts and archiving with advices causes a significant slowdown. To reduce it, we inlined the corresponding pointcuts (e.g. `modFirst`) and the method `doArchiving` by moving them to the body of set-methods. This resulted with decrease of the time overhead to the factor 5. We have experimented also with different collections and used single linked lists to implement stacks. This has resulted in a further reduction of the time overhead. However, we observed that the garbage collection is started more often in this case and in a long run there is no real gain.

# 6   Conclusion

The strong exception safety guarantee is a highly desirable property. It is equivalent to the so-called commit or rollback property. There exist patterns ensuring its satisfaction; however, they are not always applicable. In this paper, we presented an algorithm implemented in AspectJ which instruments code in a way ensuring the commit or rollback property for arbitrary methods. This algorithm is based on our earlier results concerning the computation of old attribute values. We discussed also the results concerning time- and space-overhead.

# References

1. Abrahams, D.: Exception-safety in generic components—lessons learned from specifying exception-safety for the C++ standard library. In: Jazayeri, M., Loos, R., Musser, D.R. (eds.) Generic Programming, LNCS, vol. 1766, pp. 69–79 (1998)
2. Alexandrescu, A., Held, D.: Smart pointers reloaded (ii): Exception safety analysis. C/C++ Users J. **21**(12), 40 (2003)
3. Austern, M.: Standard library exception policy. C++ Standards Committee Papers (1997). http://www.open-std.org/jtc1/sc22/wg21/docs/papers/1997/N1077.asc
4. Baar, T., Chiorean, D., Correa, A., Gogolla, M., Hussmann, H., Patrascoiu, O., Schmitt, P., Warmer, J.: Tool support for OCL and related formalisms—needs and trends. In: Bruel, J.M. (ed.) Satellite Events at the MoDELS 2005 Conference, LNCS, vol. 3844 (2006)
5. Beck, K.: Smalltalk Best Practice Patterns. Prentice Hall (1997)
6. Barnett, M., Leino, R.K.M., Schulte, W.: The Spec# programming system: an overview. In: CASSIS 2004, LNCS, vol. 3362 (2004)
7. Conchon, S., Fillitre, J.C.: Semi-Persistent Data structures. In: Drossopoulou, S. (ed.) ESOP 2008, LNCS, vol. 4960, pp. 322–336 (2008)
8. Driscoll, J.R., Sarnak, N., Sleator, D.D., Tarjan, R.E.: Making data structures persistent. J. Comput. Syst. Sci. **38**(1) (1989)
9. Dzidek, W., Briand, L., Labiche, Y.: Lessons learned from developing a dynamic OCL constraint enforcement tool for java. In: Best Papers of Satellite Workshops at the Models'05 Conference, LNCS, vol. 3844, pp. 9–19 (2006)
10. Darvas, A., Müller, P.: Reasoning about method calls in JML specifications. In: Proceedings of the 7th Workshop on Formal Techniques for Java-like Programs (FTfJP'05), Glasgow, Scotland (2005)
11. Gray, J., Reuter, A.: Transaction Processing: Concepts and Techniques. Morgan Kaufmann (1993)
12. Griffiths, A.: More Exceptional Java (2002). http://www.octopull.demon.co.uk/java/MoreExceptionalJava.html
13. Hussmann, H., Finger, F., Wiebicke, R.: Using previous property values in OCL postconditions: an implementation perspective. In: International Workshop 'UML 2.0—The Future of the UML Constraint Language OCL', 2'nd of October, York, UK (2000)
14. Kosiuczenko, P.: On the implementation of @pre. In: Chechik, M., Wirsing, M. (eds) Fundamental Approaches to Software Engineering, LNCS, vol. 5503, pp. 246–261 (2009)
15. Kosiuczenko, P.: An abstract machine for the old value retrieval. In: Bolduc, C., Desharnais, J., Ktari, B. (eds.) Mathematics of Program Construction (MPC 2010), LNCS, vol. 6120. Springer (2010)
16. Meyer, B.: Applying design by contract. Computer, vol. 25(10), pp. 40–51. IEEE Computer Society Press (1992)
17. Meyer, B.: Eiffel: The Language. Object-Oriented Series. Prentice Hall (1992)
18. Munkby, G., Schupp, S.: Automating exception-safety classification. Science of Computer Programming, vol. 76(4). Elsevier (2011)
19. OMG: OCL Specification, Version 2.0, Formal/2006-05-01 (2006)
20. Stroustrup, B.: Exception safety: concepts and techniques. Advances in Exception Handling Techniques, LNCS, vol. 2022, pp. 60–76. Springer (2001)
21. Vogel, L.: Eclipse Test and Performance Tools Platform (TPTP). http://www.vogella.de/articles/EclipseTPTP/article.html

# Efficient Testing of Time-Dependent, Asynchronous Code

**Tomasz Lewowski**

**Abstract** Time is important. New software rarely works as a standalone service—usually, it is part of a much bigger network of interconnected applications, running on different hardware, operating systems and designed with various technology stacks. Even applications which do not need network connectivity, such as some embedded devices, rely on time-based events for task scheduling. However, support for testing time-dependent code is often not sufficient both on operating system level and on framework/library level, and generally, requires special ways of designing software to provide sufficient testability. Custom design often means implementing special testing library or modifying the source code for test purposes. This paper points out some of the things that may go wrong when using these approaches, introduces a portable model of time-based computation and presents zurvan—an open-source library for Node.js applications which provides an abstraction over time and allows users to efficiently test time-dependent, asynchronous code. Advantages of using presented model include test case repeatability, increased code testability and shorter test suite execution times.

## 1 Introduction

Testing time-dependent code is hard. In functional programming paradigm, time is treated as side effect and abstracted. This paper proposes a similar approach—to abstract real-world time[1] consumption from the program functionality. Time consumption depends on many factors, which are often irrelevant for the development team—these include used hardware and its configuration, other processes running

---

[1]Real-world time is a phrase used in this paper to differentiate time passing in the real world from simulated time, chosen so that it won't collide with the term real-time, used for real-time systems.

---

T. Lewowski (✉)
Faculty of Computer Science and Management, Department of Software Engineering,
Wrocław University of Science and Technology, Wrocław, Poland
e-mail: tomasz.lewowski@pwr.edu.pl

© Springer International Publishing AG 2018
P. Kosiuczenko and L. Madeyski (eds.), *Towards a Synergistic Combination of Research and Practice in Software Engineering*, Studies in Computational Intelligence 733, DOI 10.1007/978-3-319-65208-5_6

on same machine and operating system scheduler configuration. Still, software has to be tested in some conditions, on some hardware, with some other processes and some system scheduler.

Abstraction over time is not a new subject—other papers, such as [21] use this method to increase software testability and, consequently, quality. However, these papers do not present exact model of abstraction over time.

This paper introduces a model for time-based computation. It also presents *zurvan* [20], a Node.js library for managing event queue which implements presented model and is already used in real-world software development. The model can be also implemented in other software development environments (assuming that they deal with events via a message queue).

Example code is written in JavaScript [10] and may require some knowledge about Promises [16] and JavaScript syntax. This language was selected due to its ubiquity and the fact that example implementation of the model is written in JavaScript and available for Node.js platform [25].

The rest of paper is organized as follows: Sect. 2 describes in detail why testing of time-dependent code is difficult, Sect. 3 presents a simple, yet powerful model of time-based computation, Sect. 4 contains detailed information about *zurvan*, a library that implements presented model. This library was developed as a result of special needs in the industry and slowly becomes more widely adopted. Section 5 mentions possible problems with widespread adoption of the model, Sect. 6 describes plans for further development and Sect. 7 concludes the paper with a short summary of its contributions.

## 2  Problems with Time-Based Code

Verification of asynchronous code is a problem studied by many researchers [4, 8, 9, 14, 17, 19]. This work mostly focuses on providing higher quality guarantees [13] and on multithreaded environments [7].

On the other hand, the problem of time dependence is often ignored, under the assumption that time can be treated as just another source of events [6]. While this is a valid assumption, time is a very specific source which obeys several rules, such as monotonicity. By using these rules (described in Sect. 3) it is possible to create a model of time-based computation which can be useful for many projects.

There are several important challenges related to testing asynchronous and time-based code:

1. performance—tests (especially unit tests) should be as fast as possible, while dependence on real-world time puts lower bounds on execution times,
2. non-determinism—external processes, not relevant for production environment, may require processor time and cause intermittent failures of test cases,
3. hidden hooks—it is much harder (comparing to synchronous code) for testers to check if the asynchronous system has done its processing or is there something

more that will trigger additional calculations (particularly important in multi-threaded and distributed systems).

Software systems which rely on time are generally harder to test than ones that do not. While the same applies to all side effects—including disk I/O and network connections—time is arguably one of the hardest side effect to avoid. Typical stubbing does not work well for time-based events, because of certain invariants that have to be preserved by a sane time source, discussed later in the paper.

## 2.1   Performance

Performance may be a significant problem in software systems which are time-dependent. A common problem during testing of software—especially legacy software—is testing failure scenarios—for example, communication timeouts. In an application which was not well-designed, waiting for predefined amount of time is often the only available solution. This significantly impairs software testability, as this type of tests can take a lot of time. It is also common for this type of applications not to be able to run tests in parallel (for instance, due to database dependencies). These two facts, when analyzed together, render testing of such scenarios virtually impossible, thus they are often not tested in legacy software (some authors go even as far to define the term *legacy software* as *software without tests* [12]). And software which is not tested is likely to contain more defects than tested one.

## 2.2   Non-determinism

It is well-known that code with side effects is generally harder to test than the one without them. Time dependence can be treated as a side effect—one cannot assume how long will the action take in test environment, even if time frames are exactly specified for production environments. There are simply too many test environments—while it is possible to enforce certain performance characteristics in User Acceptance Testing environment (which should be as close to production one as possible), and sometimes even in Quality Assurance departments, it is virtually impossible to provide consistent set of characteristics for developers' machines in unit tests and low-level integration tests. Dependence on time causes non-determinism, which may be a reason for intermittently failing test cases.

Intermittently failing test cases are very troublesome for software development. While every developer shall assume that a test case failure means either a bug in code or wrong test case, this is much harder for time-dependent code. When failure is not reproducible in different environments (for example, with more processing power), the problem may be attributed to machine configuration instead of an actual race condition.

Psychologically, when a stimuli (failed build) is received often and does not require special attention (fail is intermittent and caused by something which is out of development team interest, such as network connectivity, domain name resolution or system scheduling), every further occurrence of the same stimuli will cause less response, leading to ignoring actual defects. This effect is known in psychology as habituation [23]. This is especially true for tests in multithreaded environments, where real defects may be intermittent and very hard to reproduce.

## 2.3 Hidden Hooks

In multithreaded code it is not always obvious when does entire processing end. Waiting for certain events is a typical approach, but it ignores the fact that additional, unwanted computation may be performed after all valid events are received and test case ends with success (in some cases events from test A may even cause test B to fail). If code depends on time, this becomes even more troublesome, as tester cannot even wait until all events are processed, since new ones will appear all the time (for example, if the system includes any form of polling).

## 3  A Model of Time-Based Computation

Time scheduling is a complex subject, thoroughly studied in the area of operating systems. However, most of the applications do not—and shall not—rely on the used scheduling algorithm. This is particularly important when discussing portable software, which can be executed on multiple operating systems—such software cannot rely on specific scheduling algorithms.

For simplicity the model outlined below is presented for a single threaded application handling messages in FIFO mode from an inspectable message queue (i.e., one that can be synchronously queried for number of awaiting messages). In multithreaded applications it becomes slightly more complicated, since synchronization between threads is needed as well—however, same basic principles may be applied. Understandably, on process layer synchronization becomes more complicated again.

Approach in this paper assumes that a thread in every point of time can be classified as one of three states:

1. processing (thread busy),
2. idle (no processing—new tasks may be assigned),
3. blocked (a long computation takes place or I/O action is ongoing, thread busy).

This paper postulates that by simulating an *infinitely fast processor* and possibly interspersing its processing with blocking actions one can obtain a useful model for further computation and basic functional tests.

## 3.1  Infinitely Fast Processor

Infinitely fast processor can be thought of as one which performs all actions performable at given point of time before next time slice. While this requirement may seem obvious, it is not trivial in most real-life applications. This is because it imposes a significant constraint on the interaction with real world—it becomes important whether file read is instant, takes 100 ms or times out after 5 min. This is another issue that has to be analyzed by software designers and taken into account by testers. While this might seem like an additional design overhead, it can also contribute to the general quality of software by motivating designers to ask questions like *what will happen if our I/O actions will not be executed in the typical order?*

## 3.2  Action Store

When processing of all events for a given point of time finishes, system becomes idle to the next important point of time. *Important* means a point at which next action is triggered. This action can be a freshly established network connection, file read callback, cyclic event or any action triggered by the test driver. It is important to note that test driver must be aware of all the events that may happen and take time—this usually means that test driver either terminates all these events (acts as a mock or stub), serves as a proxy, or at least is notified about them. This is quite an unfortunate fact, since it makes the test driver aware of the state of the whole application environment, often not deemed important for currently tested feature. Still, this approach very much resembles the one taken by several pure functional languages—every side effect (network connection, file read) remains a side effect, regardless of its perceived importance.

## 3.3  Forwarding Procedure

When all events scheduled at current time slice are processed, the time comes to advance to the next important time slice. However, from tester perspective this is cumbersome—test cases are not written per *time slice* but per feature. If the system uses one type of events occurring every 1 s and second type, occurring every 15 s, and tester writes test case for an event of second type, some support from the tooling is needed. It would be wrong to advance whole 15 s at once, since there are 15 events that should occur in the meantime and functionality may depend on it. At the same time it would be cumbersome to 15 times advance time slice manually.

For that reason a procedure for forwarding time is needed. The procedure is described in Algorithm 1.

---

**Data**: final timestamp
**while** *current timestamp ≤ final timestamp* **do**
  perform all actions in current time slice
  increment timestamp
  advance to next time slice
**end**

---

**Algorithm 1:** Time forwarding procedure

It is important to note that actions in current time slice will often schedule other actions in same time slice (for example, send messages to other components), thus the actual implementation becomes much more complex.

## 4 Real-Life Implementation

*zurvan* [20] is a library designed to aid development teams in writing tests for time-dependent code without performance penalties and with deterministic expiration times. It has two main features: first, it is an event queue manager, allowing the tester to wait until all effects of test scenario are known, and second, it keeps track of time-based events. While time-based events can be also tracked by other solutions (like *lolex* [1]), they do not provide the necessary event queue management. *zurvan* is written in JavaScript (ECMAScript 5 [10]) and compatible with all Node.js releases starting from 4.0. It can be used in every language which uses Node.js platform (although writing language-specific bindings may be needed). It is available as an open source library on GitHub [15] and as a package on npm [22].

### 4.1 Design

*zurvan* is designed to utilize native `Promise` interface of Node.js. To provide best support for natural interaction with Node.js platform, most API functions are asynchronous and return `Promise` (with notable exceptions of `zurvan.block System(...)` and `zurvan.setSystemTime(...)`). Core concept of zurvan is the event loop. zurvan intercepts timer calls (`setTimeout`, `setInterval`, `setImmediate`), but not native Promise calls or process. nextTick calls, which all are basically a way to delay the computation until next tick of the event loop. zurvan assumes that there can be only a fixed-length chain of `setImmediate` in every single tick of the event loop, and it also assumes that `setImmediate` operates on the macroqueue. While Node.js does not exactly follow the notion of microqueue and macroqueue, this assumption is approximate enough to make *zurvan* work in most real-life scenarios. For these rare cases where this is not sufficient, *zurvan* offers a configuration option that increases the chain of `setImmediate` assumed to form a single event loop cycle.

## 4.2   Utilities and Configuration

To provide better developer experience *zurvan* has several baked-in features, which
are not strictly required for the time model, but nevertheless make development easier. These functionalities include a simple `TimeUnit` class which supports basic
operations (adding/subtracting/conversion) for managing timeout times, option to
forward time to expire all timeouts which are set, forwarding only to next timer
and a plethora of configuration options for enabling various configuration checks,
such as forbidding too long timeout times or accepting only functions as callbacks
(not strings which would need to be evaluated later on). *zurvan* also allows user to
intercept `Date` object (system time) and timers managed by process (`process.`
`uptime()` and `process.hrtime()`).

## 4.3   Usage

An example of *zurvan* usage is presented in Listing 1. The code, when executed,
asynchronously creates an array (`results`) which contains elements: `[1, 2, 3,`
`4, 5, 4]`. Elements are added at timestamps 0 (1, 2, 3—Node built-in `Promise.`
`resolve, process.nextTick` and `setImmediate`), 50 ms (4—first result
of `setInterval` call) and 100 ms (result of `setTimeout` and second result
of `setInterval`). All used scheduling functions (`setImmediate, Promise.`
`resolve, setInterval, process.nextTick` and `setTimeout`) are built-
in Node.js functions (some are defined by ECMAScript standard, others are Node.js
specific).

There are four interactions with *zurvan*—first at line 4. At this internal Node.js
timers are substituted by simulated ones provided by *zurvan*. If any calls to scheduling functions are made before `zurvan.interceptTimers()` they will use
original timers and real-world time. Second call, at line 23 executes all actions that
are awaiting on the event queue—this includes `Promise.resolve, Promise.`
`reject, process.nextTick, setImmediate` and timers which expire at this
point of time (e.g. `setTimeout(f, 0)`). Third, at line 29 the code simulates
time flow—timestamp is forwarded to closest event (`setImmediate` at timestamp
50 ms), the event is executed and forwarding continues up to the requested time
(100 ms). Finally, the call at line 32 releases all acquired resources (timers, `Date`
object) and returns all remaining timers to the caller—this feature may be used, for
example, to assert that all scheduled events were executed in the test case. *zurvan* is
asynchronous by default, so its calls generally return instances of `Promise`, which
are resolved at specific points of time, after the event queue is emptied.

**Listing 1**  Example of *zurvan* usage

```
1    var zurvan = require('zurvan');
2    var results = [];
3    zurvan.interceptTimers()
4      .then(function() {
5        setTimeout(function() {
6          results.push(5);
7        }, 100);
8
9        setImmediate(function() {
10         results.push(3);
11       });
12
13       process.nextTick(function() {
14         results.push(2);
15       });
16
17       Promise.resolve()
18         .then(function() {
19         results.push(1);
20         });
21
22       return zurvan.waitForEmptyQueue();
23     })
24     .then(function() {
25       setInterval(function() {
26         results.push(4);
27       }, 50);
28       return zurvan.advanceTime(100);
29     })
30     .then(function() {
31       return zurvan.releaseTimers();
32     })
```

A more complex example of *zurvan* usage is presented in Listing 2. The code in Listing 2 creates an array (`results`) which—after the code finishes execution—will contain elements: `[1, 2, 1]`. This might seem confusing, since number 1 should be pushed every 50 ms (scheduled at line 9) and 150 ms should have elapsed (time forwarding issued at line 12). However, after 75 ms a blocking action is simulated (line 7), which prevents callbacks from being executed immediately, like a real blocking action. This is why callbacks are executed at timestamps: 50, 75 and 175 ms.

**Listing 2** Example of usage of both blocking and forwarding actions

```
1   var zurvan = require('zurvan');
2   var results = [];
3   zurvan.interceptTimers()
4     .then(function() {
5       setTimeout(function() {
6         results.push(2);
7         zurvan.blockSystem(100);
8       }, 75);
9       setInterval(function() {
10        results.push(1);
11      }, 50);
12      return zurvan.advanceTime(150);
13    })
14    .then(function() {
15      return zurvan.releaseTimers();
16    });
```

## 4.4 Differences from Other Solutions

zurvan is not the only library which attempts to solve problem of timing events in Node.js. The most well-known and popular solution as of today is *lolex*, part of *sinonjs* project. Both *lolex* and *zurvan* provide fake implementations of set Timeout, setInterval, setImmediate, clearTimeout, clear Interval and clearImmediate, simulated Date instances and simulated process.hrtime/process.uptime. The most important difference between them is how they handle asynchronous actions—*lolex* assumes all actions are either synchronous or depend on one of the timers (usually setImmediate). *zurvan* makes no such assumptions and delivers forwarded time as a Promise. While this causes *zurvan* to be a more complex and less beginner-friendly solution, it also allows users to freely use Node.js features such as native Promise implementation or process.nextTick, which is not possible when using *lolex*.

## 4.5 Limitations

zurvan is a library designed specifically to handle timing events. This means that it handles only timers and does not care about other asynchronous actions, such as filesystem operations, network I/O or inter-process communication. Another limitation is that it performs function substitution on JavaScript level - this has two major consequences (none of them should be a major drawback for a well-designed application, but the developer needs to be aware of them):

1. real timers may be captured to a local variable by some global initialization code,
2. code executed from external modules uses real-world time.

First one means that it is not sufficient to `require` the package to use it. Timers (and other functions) are only substituted between calls to `zurvan.intercept Timers` and `zurvan.releaseTimers`. This means that all objects created before timer interception may have access to original timers and use them via closures. This is particularly important for packages which have a global initialization routine, which may cache some variables.

Second one means that code included from modules written in C++ still can use original timers. This will of course render whole model useless, but using C++ modules for triggering asynchronous actions is rather rare in practice, so this does not appear to be a major threat.

Additional failure scenario which would work in real life is scenario of busy wait on the event queue. Code such as one in Listing 3 will eventually end in real environment, but will fail with *zurvan* due to infinitely fast processor approximation. Arguably, busy wait is rarely the best design possible, so it should not be a major drawback. Listing 3 presents an example of such a busy wait.

**Listing 3** Infinite loop in *zurvan* but not in real world

```
1  var uptime = process.uptime();
2  var t = function() {
3    setImmediate(function() {
4      if(process.uptime() < uptime + 2) {
5        t();
6      }
7    })
8  };
9  t();
```

It is important to note that all these failure scenarios are not *zurvan*-specific, but apply to its competitors' solutions as well.

## 4.6 Quality

*zurvan* was designed from the beginning to be delivered with quality in mind. Because of that its test cases cover 100% of statements and over 95% of branches. Tests are executed using hospitality of Travis CI [26] and static analysis is done using CodeClimate [3] tools.

## 5 Threats to Validity

Presented model was validated on a large-scale project (over 100 developers) in Nokia and was proven useful in industrial practice. *zurvan* itself is used by several companies and has no open bug reports as of today.[2] However, it is necessary to remember that presented model aims to provide maximum reproducibility of test runs, not the highest flexibility of scenarios. It is possible that—for particular implementations—some defect scenarios will either not be reproducible or will be very hard to reproduce (still, should not be harder than in a test case without simulated time). This is an inherent limitation of adopting a time model such as one presented in the paper.

Tests basing on presented time model do not aim to provide all possible interspersing sequences—this is not possible via testing due to computational effort and usually requires model checkers [2] or other formal tools.

## 6 Further Work

Current work on *zurvan* focuses on enabling it to work on WebKit engine ([27]) in the browser, which is the main goal for 1.0 release. For Node.js development it would be beneficial to provide a complex solution which would substitute not only timers, but filesystem and network access as well. In such a testing environment tests would become much more reliable and less dependent on the environment.

A lot is to be done for testing time-dependent code, especially in a cross-language manner. While it is possible to design the application in such a way that time-dependent parts are extracted to separate module which is mocked during tests, it requires conscious effort from the development team. It is obviously not possible to easily introduce such design into existing legacy code. For that reason, a promising research area is providing simulated time on lower level. This can be evaluated on core library level, operating system level or virtual machine level.

A low-level approach may yield many more challenges than implementation on library level (such as *zurvan*), but at the same time covers many more use cases. Still, until such a system is ready, implementations on library level are needed in other languages as well. In languages with a notion of thread—such as C++ [5] or Java [18]—such an implementation will probably be needed per each thread-wrapping library, as it will rely on thread message/task queues, which is custom in every library. In languages which rely on actors, such as Scala [24] or Erlang [11], a thin wrapper around event queue may be sufficient.

---

[2]28 May 2017.

# 7 Conclusions

This paper introduced a model of time-based computation, together with a sample implementation for Node.js, available as an open source library and used in the industry. Testing asynchronous, time-dependent code is never easy, but this paper has laid foundation for implementation of simulated time for test purposes on various level of software system. Using simulated time does not aim to allow to test all possible task intersperse scenarios, but only a basic subset of them. Yet, industrial practice has proven that such solutions are needed, since the open-source libraries implementing some flavor of presented model (*lolex* and *zurvan*) are either already adopted or getting adoption in the industry.

# References

1. Antoni, M., Tomlinson, C., Goldberg, J., Kopsend, C.E., Roderick, M., Edwards, A.: lolex. https://github.com/sinonjs/lolex (2017)
2. Bérard, B., Bidoit, M., Finkel, A., Laroussinie, F., Petit, A., Petrucci, L., Schnoebelen, P.: Systems and Software Verification: Model-Checking Techniques and Tools. Springer Science & Business Media (2013)
3. CodeClimate. https://codeclimate.com/ (2017)
4. Corrodi, C., Heußner, A., Poskitt, C.M.: A Graph-Based Semantics Workbench for Concurrent Asynchronous Programs, pp. 31–48. Springer, Berlin, Heidelberg (2016)
5. C++. https://isocpp.org/ (2017)
6. Czaplicki, E., Chong, S.: Asynchronous functional reactive programming for guis. SIGPLAN Not. **48**(6), 411–422 (2013). doi:10.1145/2499370.2462161
7. Deligiannis, P., Donaldson, A.F., Ketema, J., Lal, A., Thomson, P.: Asynchronous programming, analysis and testing with state machines. SIGPLAN Not. **50**(6), 154–164 (2015). doi:10.1145/2813885.2737996
8. Desai, A., Garg, P., Madhusudan, P.: Natural proofs for asynchronous programs using almost-synchronous reductions. SIGPLAN Not. **49**(10), 709–725 (2014). doi:10.1145/2714064.2660211
9. Desai, A., Qadeer, S., Seshia, S.A.: Systematic testing of asynchronous reactive systems. In: Proceedings of the 2015 10th Joint Meeting on Foundations of Software Engineering, ESEC/FSE 2015, pp. 73–83. ACM, New York, NY, USA (2015). doi:10.1145/2786805.2786861
10. ECMAScript Language Specification. https://www.ecma-international.org/ecma-262/5.1/ (2011)
11. Erlang. https://www.erlang.org/ (2017)
12. Feathers, M.: Working Effectively with Legacy Code. Prentice Hall Professional (2004)
13. Ganty, P., Majumdar, R.: Algorithmic verification of asynchronous programs. ACM Trans. Program. Lang. Syst. **34**(1), 6:1–6:48 (2012). doi:10.1145/2160910.2160915
14. Garavel, H., Lang, F., Mateescu, R.: Compositional verification of asynchronous concurrent systems using cadp. Acta Informatica **52**(4), 337–392 (2015). doi:10.1007/s00236-015-0226-1
15. GitHub. https://github.com/ (2017)
16. Hussain, M.: Mastering JavaScript Promises. Packt Publishing Ltd (2015)
17. Isabel, M.: Testing of concurrent programs. In: Carro, M., King, A., Saeedloei, N., Vos, M.D. (eds.) Technical Communications of the 32nd International Conference on Logic Programming (ICLP 2016). OpenAccess Series in Informatics (OASIcs), vol. 52, pp. 1–5. Schloss

Dagstuhl–Leibniz-Zentrum fuer Informatik, Dagstuhl, Germany (2016). doi:10.4230/OASIcs. ICLP.2016.18. http://drops.dagstuhl.de/opus/volltexte/2016/6747
18. Java. https://www.java.com/ (2017)
19. Lei, Y., Tai, K.C.: Efficient reachability testing of asynchronous message-passing programs. In : Eighth IEEE International Conference on Engineering of Complex Computer Systems, 2002. Proceedings., pp. 35–44 (2002). doi:10.1109/ICECCS.2002.1181496
20. Lewowski, T.: zurvan. https://github.com/tlewowski/zurvan (2017)
21. López, M., Castro, L.M., Cabrero, D.: Feasibility of Property-Based Testing for Time-Dependent Systems, pp. 527–535. Springer Berlin, Heidelberg, (2013)
22. npm. https://www.npmjs.com/ (2017)
23. Rankin, C.H., Abrams, T., Barry, R.J., Bhatnagar, S., Clayton, D.F., Colombo, J., Coppola, G., Geyer, M.A., Glanzman, D.L., Marsland, S., McSweeney, F.K., Wilson, D.A., Wu, C.F., Thompson, R.F.: Habituation revisited: an updated and revised description of the behavioral characteristics of habituation. Neurobiol. Learn. Mem. 92(2), 135–138 (2009). doi:10.1016/j.nlm.2008.09.012. http://www.sciencedirect.com/science/article/pii/S1074742708001792. Special Issue: Neurobiology of Habituation
24. Scala. http://scala-lang.org/ (2017)
25. Tilkov, S., Vinoski, S.: Node.js: using javascript to build high-performance network programs. IEEE Internet Comput. 14(6), 80–83 (2010). doi:10.1109/MIC.2010.145
26. Travis CI. https://travis-ci.org/ (2017)
27. WebKit. https://webkit.org/ (2017)

# Automatic Processing of Dynamic Business Rules Written in a Controlled Natural Language

**Bogumiła Hnatkowska and Tomasz Gawęda**

**Abstract** Business rules are such requirements that can change very often. As they are formulated by business people (e.g. domain experts) they should be expressed in the way that is—from one side—easy to understood and—from the other—possible for automatic processing. This paper demonstrates a solution to the processing of dynamic business rules which are written in a controlled natural language. A user can add or modify rules during program operation influencing the way the program behaves. The proof-of-concept implementation confirmed that such approach is feasible and can be extended to become mature enough to be introduced in production.

**Keywords** Dynamic business rules · Action enablers · Pre-conditions · Post-conditions · Controlled natural language · SBVR

## 1 Introduction

Business rules belong to the most significant output artifacts of business analysis and are valuable citizens of requirements world [1]. As they are under the business jurisdiction, they are often subjects to change. Unfortunately, such changes very often result in re-implementation of a program logic.

There exist rule engines, like JBoss Drools [2], that enable partial separation of business rules and application logic. However, they do not support expressing business rules in natural language, which is the most appropriate for domain experts [1]. What is more, the connection between application business logic and rules

B. Hnatkowska (✉) · T. Gawęda
Faculty of Computer Science and Management, Wroclaw University of Science
and Technology, Wroclaw, Poland
e-mail: Bogumila.Hnatkowska@pwr.edu.pl

T. Gawęda
e-mail: tomasz.gaweda@outlook.com

© Springer International Publishing AG 2018
P. Kosiuczenko and L. Madeyski (eds.), *Towards a Synergistic Combination
of Research and Practice in Software Engineering*, Studies in Computational
Intelligence 733, DOI 10.1007/978-3-319-65208-5_7

stored by the engine is strict as programmers call rules directly from application code or application code can be run from a particular rule.

There were successful attempts to solve the problems mentioned above with the usage of Aspect Oriented Programming (AOP), e.g. [3, 4], in which a domain specific language (DSL) was proposed to define activation events, describing moments when to evaluate selected business rules. Business rules in this solution were expressed in DRL language, supported by JBoss Drools.

This paper presents a new approach to implementing an integration layer between application business logic and rule engine. In this approach business rules are written in a controlled natural language similar to SBVRSE [5]. The connections between business logic and business rules are described in a declarative manner with the use of Java annotations, and served by Spring AOP. Business rules are automatically translated to MVEL dialect of JBoss—an expression language for Java-based applications.

The rest of the paper is structured as follows. Section 2 brings short characteristics of dynamic business rules together with the way they can be expressed in semi-natural language. Section 3 presents the proposed solution to implement integration layer, responsible for business rules processing. A representative case study for which the solution was applied is given in Sect. 4. The last Sect. 5 summarizes the paper and presents directions of further works.

## 2 Dynamic Business Rules and Their Representation in Natural Language

In general business rules are classified into two mains groups: static (also called definitional) and dynamic (also known as operative) [5]. Static business rules are easier to be maintained as they refer to entities, their properties, and relationships among them. For example, one can state "Each student must have a unique index number" or "Each order must have a shipping date that precedes the order date". These requirements very often represent invariants which can be checked in many application layers, e.g. on the client-side, on the server-side or by a database. There are many approaches, e.g. [6, 7], which address the transformation of definitional business rules into a programming language. However, there are only few related to dynamic business rules. One example is [8] but here the targeted language is OCL [9]—functional associate of UML. The other is [10] where the authors present AODA framework in which post-conditions can be expressed either in OCL or in Drools DSL, and on that basis translated to Java service annotations. Unfortunately, [10] does not address business rules expressed in natural language.

The paper focuses on dynamic business rules, especially action enablers and action pre, and post-conditions. The intended behavior of the application using above mentioned types of business rules is that an action enabler runs connected action in predefined order. Pre/post conditions throw an exception if the condition is not met.

Business rules can be written using different formalisms, e.g. OCL, JDrools DSL, however, according to Business Rule Manifesto, they should be expressed in natural language to enable effective communication among business experts and software developers [1]. Unlackily, natural language has its deficiencies, e.g. can be misinterpreted or incomplete. It is also difficult to process sentences written in natural language automatically by tools. A compromise solution is to apply a controlled natural language for this purpose. The syntax of the controlled natural language is limited, but it can be checked and automatically interpreted.

Semantics of Business Vocabulary and Business Rules [5] is an OMG standard "for the exchange of business vocabularies and business rules among organizations and between software tools". It proposes two dialects of controlled natural language: SBVR Structure English (SBVRSE), and the RuleSpeak business rules notation. Because of its popularity, SBVRSE was adapted to express dynamic business in this work, and its way to express obligation statements ("It is obligatory ..."), and prohibition statements ("It is prohibited ...").

The original syntax of SBVRSE was a little bit extended to enable direct definition of when to check a condition (before an action or after it) or when to run enabling action (before enabler action or after it). An exemplary pre-condition is given below:

"If a customer rent a car, it is obligatory that the car must be clean".

Instead of "if" one can use "when" or "before" (the semantics is the same) or "after"—now the condition ("car must be clean") will be checked after "rent car" action. The grammar mistake in the example ("customer rent") is intentional—it is a side-effect of application of controlled natural language which authors wanted to be easy to parse and interpret.

It should be noted that dynamic business rules are built on facts and facts are built on concepts as expressed by terms [1]. Both facts and terms are elements of business vocabulary. In consequence, a tool supporting definition of dynamic rules must also know used vocabulary.

# 3 Architecture of Proposed Solution

## 3.1 Workflow Description and Functional Architecture of the Solution

Business rules processing requires three sub-processes to be run in a sequence (see Fig. 1):

1. Definition of business vocabulary.
2. Definition of business rules.
3. Processing of business rules.

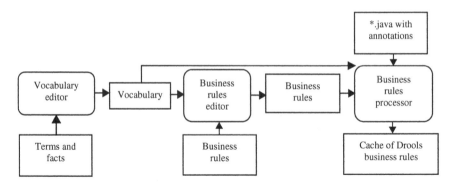

**Fig. 1** Business rules processing workflow

**Fig. 2** Functional
architecture of proposed
solution

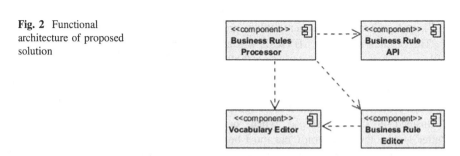

Rounded rectangles represent processes, normal rectangles—input/output data,
and arrows—data flow.

Particular sub-processes are realized as separate components presented in Fig. 2.
There is also an additional project involved, called Business Rule API, which
defines elements, e.g. annotations. Detail description of the main components is
given in the following subchapters.

## 3.2 Vocabulary Editor

Terms and facts are main elements of business vocabulary [5]. For their representation, a simple grammar was defined with the use of Xtext [11]. The tool
automatically provides an appropriate parser and generates an editor with colored
syntax and contextual hints. It was assured that facts can refer only to existing
terms. The vocabulary editor can be installed as an Eclipse plugin.

An example of vocabulary definition is presented in Listing 1.

```
package pl.edu.pwr.example {
  Term: customer
    Description: 'Driver that bought something.'
```

```
Term: order
Term: car
Fact: customer places order
Fact: customer rent car
}
```

Listing 1. Example of business vocabulary

We expect defined terms and facts be represented by Java classes included in one package (pointed out in the first row). In this example, we have three terms and two facts defined. A term or fact must be represented by one word, but it can include '_' symbol. Now only binary facts are supported.

## 3.3 Business Rules Editor

Business rules editor was also prepared with Xtext. The first element in its construction was to define a formal grammar to represent business rules. As it was mentioned in Sect. 2, we limit our interests to action enablers and pre/post conditions for actions. In consequence, we had to decide about possible forms of conditions. Now two cases are covered:

- Terms comparison—a term typically represents a whole class or its attribute. Two terms can be compared with the use of built-in operators ('is equal', '>', '<', 'is at least', etc.); Relational operators are implemented only for integer values. Example: If a driver rent a car, then it is obligatory that the age of the driver is at least minimumAllowedDriverAge of car.
- Comparison of a term and a literal value; term must be an integer value now. Example: If a driver rent a car, then it is obligatory that the <u>age of driver is at least 18</u>.

BNF notation for a business rule can be shortly written as:

$$< RuleName > \ < ActionPlace > \ < Clause >, (then)? \ < ModalOperator >$$
$$( \ < BinaryTestClause > \ | \ < InvocationClause > \ )$$

where:

| | |
|---|---|
| RuleName | Rule name, e.g. 'Rule MinimalLevel is' |
| ActionPlace | Where to check the rule (before action or after it) |
| Clause | Action to be run |
| ModalOperator | Obligation or prohibition statement |
| BinaryTestCase | Condition type to be checked |
| InvocationClause | Identification of another action and its parameters (if necessary) to be run in case of action enabler rule |

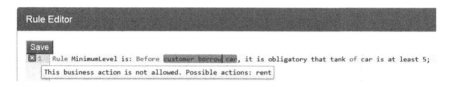

**Fig. 3** Business rule editor in action

In further two simple examples of different business rules are given:

(a) Rule MinimumLevel is: Before customer rent a car, it is obligatory that the tank of the car is at least 10.
(b) Rule CreateInvoice is: After customer rent a car, then it is obligatory that shop creates invoice using the car.

The business rules to be processed successfully require: (a) *Car* class to have *tank* attribute of *Integer* type, (b) a class with a method to create an invoice which takes a parameter of *Car* type.

Authors decided to implement business rules editor as a simple web application. Language-related services, e.g. code completion, are realized through HTTP requests sent to a server-side component. Syntax validation was implemented mostly by authors—see Fig. 3 for example.

## 3.4 Business Rules Processor

In the proposed solution JBoss Drools engine is used for evaluation of business rules. To enable this, business rules processor has first to translate business rules written in natural language to a form acceptable by Drools engine. The task is realized by *RuleMappingService*. The implementation of the service delegates the translation to a proper *MappingStrategy*. Now one strategy is implemented for obligation/prohibition statements. However, the solution, according to open-close principle, is ready for extension. Main classes being the core of rules processor are presented in Fig. 4.

Each Drools rule consists of two main sections: left-hand side (LHS) and right-hand side (RHS). LHS contains conditions which must be satisfied to run actions represented by RHS. For action enabler rule the RHS part calls another rule from rule repository with proper parameters, and for pre/post conditions, LHS contains (negated for obligations) rule condition, and RHS calls an action reporting rule violation.

An example of transformation of a business rule written in controlled natural language to the MVEL Drools dialect is given in Listing 2. Rule to be translated is given below: *CustomerHasMoney is: If customer rent car, then it is obligatory that the money of customer is at least the cost of the car.*

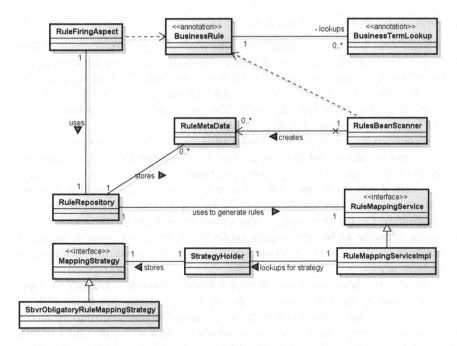

**Fig. 4** Core classes of rules processor

```
PackageDescr pkg =
DescrFactory.newPackage().name("pl.edu.pwr.gawedat")
   .attribute("dialect").value("mvel").end()
   .newImport().target("example.Car").end()
   .newImport().target("example.Customer").end()
   .newRule().name("CustomerHasMoney")
   .lhs()
     .pattern("RuleInfo").id("$ruleInfo", false)
     .constraint("businessAction=='customer rent car'")
     .constraint("when == 'before'").end()
     .pattern("Car").id("$car", false).end()
     .pattern("Customer").id("$customer", false).end()
     .eval()
     .constraint("! ($customer.money >= $car.cost)")
     .end()
   .end()
   .rhs("$ruleInfo.reportBroken('CustomerHasMoney')")
.end().getDescr();
```

Listing 2. An example of business rule in MVEL dialect

Some transformation mappings are quite obvious, e.g. the rule name is kept (.name("CustomerHasMoney")), as well as constrains describing the context, i.e. business action ("customer rent car") and the moment when to check the constraint ("if" is translated to "before"). What is more, entities involved are also imported (customer, and car). The condition itself ("the money of customer is at least the cost of the car") is represented by eval section.

Business rules translated from natural language to MVEL syntax are kept in a local cache of rules in *RuleRepository* class. This class provides API to establish a session with Drools, set up the proper context and evaluate a selected business rule.

A natural demand is that a small change in a business rule should be possible without the necessity of application to be recompiled (such a change should be transparent). On the other side, it also should be possible to add a new rule in natural language to be immediately processed if the source code is somehow annotated in advance—the annotations point out possible places in which to apply business rules.

To achieve above described behavior following solutions were applied. First, @BusinessRule annotation was introduced. The annotation has value attribute which refers to one of the facts from the vocabulary. The annotation must be placed before the methods around which we would like to check all business rules starting with "if/when/after ...", e.g. "if/when/after customer rent car". Second, @BusinessTermLookup annotation was implemented. It enables to assign method parameters with terms being parts of @BusinessRule annotation. A typical usage of both annotations is given in Listing 3. Here, *Rental* class (method parameter), has two properties: customer (referred in lookup as arg[0].customer), and car (arg [0].car). Later, instances of these terms can be passed as parameters to other action.

```
@Service
public class CarRentalService {
@BusinessRule(value = "customer rent car")
lookups = {
  @BusinessTermLookup(
      name = "customer", path = "arg[0].customer"),
  @BusinessTermLookup(
      name = "car", path = "arg[0].car") })
    public void rentCar (Rental rental) {
      // ...
}}
```

Listing 3. Exemplary application of @BusinessRule and @BusinessTerm Lookup annotations

Third, an aspect was implemented with one advice which is called around methods with @BusinessRule annotation. The aspect asks rule repository—before and after any such method is called—if all rules referring to the fact mentioned in

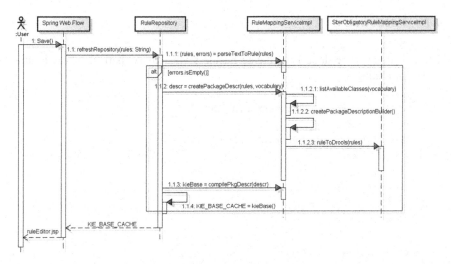

**Fig. 5** Processing of business rule change

annotation are fulfilled. If no, aspect throws an exception in order to prevent the introduction of state inconsistency.

To implement aspect part of the solution Spring AOP framework was used [12]. In consequence, projects using business rules processor do not need to be translated with AspectJ compiler.

Another advantage of Spring AOP is a possibility to use Spring application context to discover all classes marked with a @BusinessRule annotation. A dedicated class called *RulesBeanScaneer* was implemented for that purpose. This class is defined as a Spring service, and it provides one method—*scanForBusinssRules*—annotated with @PostConstruct. This way the method is called automatically after application context is created. It looks for all annotated methods and registers them in rule repository.

Business rules can be dynamically added/modified during software operation. When a user clicks save button (see Fig. 2), rules editor will call the *refreshRepository* method from *RuleRepository* class. This method: (a) checks the syntax correctness of all rules, (b) asks *RuleMappingServiceImpl* class to create JDrools package description and to translate again all rules written in the natural language to the Drools format (with the *ruleToDrools* method). See Fig. 5 for details.

## 4 Case Study

Demonstration of the proposed solution will take a form of a case-study inspired by [13]. Let assume that we have a renting car company. Before or after an event resulting from the fact that a customer rents a car, we would like to check specific conditions, among other those, checking if another action has been run.

## 4.1 *Vocabulary*

The first thing to be done is to create business vocabulary, part of which is presented in Listing 4.

```
package pl.edu.pwr.gawedat.carrental.model {
    Term: customer
        Description: 'person which have rented a car'
    Term: car
    Term: tank
        Description: 'status of tank (in litres)'
    ...
    Fact: shop own car
    Fact: shop fill tank
    Fact: shop creates invoice
    Fact: invoice has amount
    Fact: invoice isAbout rental
}
```

Listing 4. Vocabulary (part of) used in the case-study

It is expected the vocabulary to be consistent with a class model used by *Rent-Car* application. The model can be obtained automatically (see [14]) or manually. Here, it was created by hand, and is presented in Fig. 6.

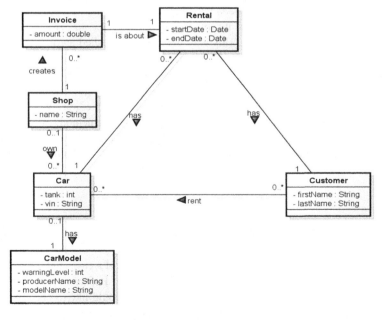

**Fig. 6** Processing of business rule change

## 4.2 Business Rules

Several business rules were defined but here we limit the presentation to four—one from each supported category. The first three serve for checking pre/post conditions while the last one is an example of action enabler (Note: the grammar mistakes are consequences of defined business vocabulary and limitations imposed by grammar definition).

1. Rule MinimumLevel is: Before customer rent a car, it is obligatory that tank of the car is at least 10;
2. Rule MinimumLevel is: After customer rent a car, it is obligatory that tank of the car is at least 10;
3. Rule MoreThanWarning is: If customer rent a car, then it is obligatory that tank of the car is greater than warningLevel of model of car;
4. Rule CreateInvoice is: After customer rent a car, then it is obligatory that shop creates invoice using the car.

## 4.3 Test Cases

A testing application was prepared with Spring Web Flow framework. In implements two flows. The first one, defined in rule.xml, has embedded the business rules editor. A user in a friendly manner can add/modify rules translated and checked by rules processor. The second one, defined in index.xml, is used for testing purposes. A user opens the page, enters customer's data, selects a car from the list and confirms his/her selection with Save button—see Fig. 7.

The testing application does not provide full rental functionality. Instead a mock of saving service is delivered which is called in reaction to *Save* button. This mock service is annotated with @BusinessRule, and—in consequence—business rule processor is called by *FireningAspect* before and after every execution of this service. In none of business rule is violated, the service informs about success, otherwise—*BusinessRuleNotFulfilledException* is thrown, and the flow is redirected to an error page on which violated rules are shown.

The same testing application was also used to check the enabler business rule. After a user clicks on Save button, in no other rules are violated, the processing engine automatically calls another service method with proper @BusinessRule annotation (value ='shop creates invoice'). In such a case a confirmation message is written ('shop has created an invoice') in the program console.

Exemplary test cases are gathered in Table 1.

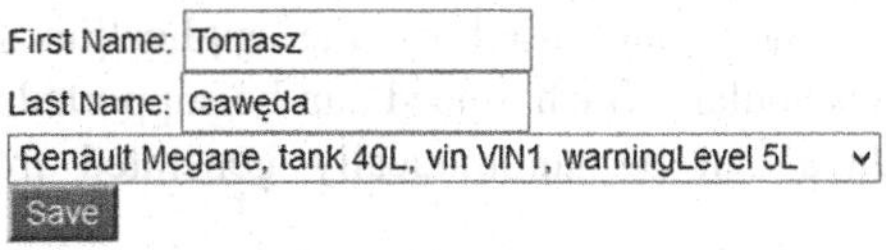

**Fig. 7** Screenshot from the testing application

First Name: | Tomasz
Last Name: | Gawęda
Renault Megane, tank 40L, vin VIN1, warningLevel 5L
Save

**Table 1** Selected test cases

| Rule number | Input data | Expected result |
|---|---|---|
| Rule 1 | Car.tank = 5 | Exception |
| | Car.tank = 15 | Valid |
| Rule 2 | Car.tank = 5 | Exception |
| | Car.tank = 15 | Valid |
| Rule 3 | Car.tank = 5 Car.model. warningLevel = 5 | Exception |
| | Car.tank = 40 Car.model. warningLevel = 5 | Valid |

## 5 Summary

Business rules are specific requirements that should be fully controlled by business people. They tend to change very often so a software system should be prepared for such change. The best-imagined solution is to enable writing business rules separately, possibly in natural language and called them when necessary in the business logic part. Application of rule engines like JBoss Drools only partially solves the problem, as the user still have to learn a domain-specific language to express business rules, and the code of application logic is tangled with the code of Drools API.

In the paper, an architecture of the solution addressing problems mentioned above is proposed. The proof-of-concept implementation confirmed its usability. Business rules, expressed in controlled natural language can be added or modified on-the-fly when the application is running. They can be run before/after business methods (for example services) that are annotated in a specific manner. A programmer can anticipate if a method is connected with business logic or not and annotate it in advance even if at that moment no business rules exist.

Similarly to [3, 4], the architecture of the processing engine applies aspect oriented programming to solve cross-cutting concern. However, in the solution, only one aspect is present that allows integrating two separate layers: application business layer and business rules processor. What is more, a user can define business rules not in a dedicated DSL but a semi-natural language with help provided by syntax highlighting and completion mechanisms.

At that moment the presented solution supports prohibition/obligation statements with only a few possibilities to define conditions. One can compare two attributes or an attribute with literal value with relational operators. The set of supported literal types should be extended as well as the possibility to define complex conditions with logical operators (and, or, not, xor).

The model used by the application should be synchronized with business vocabulary. Such a need can be supported in two different ways. Either a model in Java can be automatically generated from the vocabulary, or at least some consistency checking mechanism between Java model and defined vocabulary can be provided.

The proof-of-concept implementation was prepared using a specific technological stack: Spring AOP, JDrools, and Xtext. The solution can be easily adapted to cooperate with another business rule engine, e.g. Jess [15], which offers integration with Java language. Such migration requires implementation of a new mapping strategy from SBVR to Jess API. One can consider moving the whole solution to . NET platform as the aspect-oriented paradigm is also here available. However, such a change is more complicated as it requires rewriting business rules processor component.

# References

1. The Business Rules Manifesto. http://www.businessrulesgroup.org/brmanifesto.htm (2017)
2. Drools https://www.drools.org/ (2017)
3. Hnatkowska, B., Kasprzyk, K.: Business rules modularization with AOP. Przegląd Elektrotechniczny, R. **86**(9), 234–238 (2010)
4. Hnatkowska, B., Kasprzyk, K.: Integration of application business logic and business rules with DSL and AOP. In: Szmuc, T., Szpyrka, M., Zendulka, J. (eds.) CEE-SET 2009, pp. 30–39. Springer, Berlin (2012)
5. Semantics of Business Vocabulary and Business Rules (SBVR), vol. 1.3, OMG (2015)
6. Hnatkowska, B., Bień, S., Ceńkar, M.: Rapid application development with UML and Spring Roo. In: Borzemski, L. (eds.), Information System Architecture and Technology: Web Engineering and High-Performance Computing on Complex Environments, Oficyna Wydawnicza Politechniki Wrocławskiej, pp. 69–80, Wrocław, Poland (2012)
7. Cemus, K., Cerny, T., Donahoo, M.J.: Automated business rules transformation into a persistence layer. Proc. Comput. Sci. **62**, 312–318 (2015)
8. Bajwa, I.S., Lee, M.G.: Transformation rules for translating business rules to OCL constraints. In: SBVR vs OCL: A Comparative Analysis of Standards, 14th IEEE International Multitopic Conference (INMIC), pp. 132–143 (2011)
9. Object Constraint Language Version 2.4, OMG (2014)
10. Cemus, K., Cerny, T., Donahoo, M.J.: Automated business rules transformation into a persistence layer. Proc. Comput. Sci. **62**, 312–318 (2015)
11. Bettini, L.: Implementing Domain-Specific Languages with Xtext and Xtend. Packt Publishing (2013)
12. Aspect Oriented Programming with Spring, Pivotal Software. http://docs.spring.io/spring/docs/current/spring-framework-reference/html/aop.html (2017)
13. Concepts and Vocabulary. http://www.kdmanalytics.com/sbvr/vocabulary.pdf (2017)
14. Nemuraite, L., Skersys, T., Sukys, A., Sinkevicius, E., Ablonskis, L.: VETIS tool for editing and transforming SBVR business vocabularies and business rules into UML&OCL models. Proc. ICIST **2010**, 377–384 (2010)
15. Jess, the Rule Engine for the Java Platform. http://www.jessrules.com (2017)

# Continuous Test-Driven Development: A Preliminary Empirical Evaluation Using Agile Experimentation in Industrial Settings

Lech Madeyski and Marcin Kawalerowicz

**Abstract** Test-Driven Development (TDD) is an agile software development and design practice popularized by the eXtreme Programming methodology. Continuous Test-Driven Development (CTDD), proposed by the authors, is the recent enhancement of the TDD practice and combines TDD with the continuous testing (CT) practice that recommends background testing. Thus CTDD eliminates the need to manually execute the tests by a developer. This paper uses CTDD research to test out the idea of Agile Experimentation. It is a refined approach performing disciplined scientific research in an industrial setting. The objective of this paper is to evaluate the new CTDD practice versus the well-established TDD practice via a Single Case empirical study involving a professional developer in a real, industrial software development project employing Microsoft .NET. We found that there was a slight (4 min) drop in the mean red-to-green time (i.e., time from the moment when any of the tests fails or the project does not build to the time when the project compiles and all the tests are passing), while the size of the CTDD versus TDD effect was non-zero but small ($d - index = 0.22$). The recorded results are not conclusive but are in accordance with the intuition. By eliminating the mundane need to execute the tests we have made the developer slightly faster. If the developers that use TDD embrace CTDD it can lead to small improvements in their coding performance that, taking into account a number of developers using TDD, could lead to serious savings in the entire company or industry itself.

L. Madeyski (✉)
Faculty of Computer Science and Management, Wrocław University of Science
and Technology, Wyb.Wyspianskiego 27, 50-370 Wrocław, Poland
e-mail: Lech.Madeyski@pwr.edu.pl

M. Kawalerowicz
Faculty of Electrical Engineering, Automatic Control and Informatics,
Opole University of Technology, Ul. Sosnkowskiego 31, 45-272 Opole, Poland
e-mail: marcin@kawalerowicz.net

M. Kawalerowicz
CODEFUSION Sp. Z o.o., ul. Armii Krajowej 16/2, 45-071 Opole, Poland

© Springer International Publishing AG 2018
P. Kosiuczenko and L. Madeyski (eds.), *Towards a Synergistic Combination of Research and Practice in Software Engineering*, Studies in Computational Intelligence 733, DOI 10.1007/978-3-319-65208-5_8

# 1 Introduction

Recent surveys, case studies and white papers show wide adoption of agile methods and practices in the software industry [9, 24]. Test-Driven Development (TDD) introduced by Beck [5] is, alongside pair programming, one of the main practices in Extreme Programming (XP) which in turn is one of the main Agile software development methods. Beck in his influential book [4] shows (see Fig. 4 in Chap. 11) that both TDD and pair programming are the most interconnected practices in XP, while recent surveys and papers show increasing adoption of TDD among professional developers [1, 6, 14, 18, 22, 23].

Continuous compilation is a practice, used in modern IDEs (Integrated Development Environment), that provides source code compilation in the background that gives the developer immediate feedback about the potential compilation errors while he edits the source code. It is practically a standard in all modern IDEs like Microsoft VisualStudio (since 2010), Eclipse, IntelliJ IDEA and so on. A recent extension of continuous compilation that adds a background testing to the compilation hit the mainstream of IDEs. It is called continuous testing (CT) [20, 21]. Using CT the developer gets not only background compilation but background testing as well. CT provides an immediate test feedback on top of compilation feedback. Microsoft introduced CT in the two highest and most expensive versions of Visual Studio 2012 (Premium and Ultimate).

In our paper from 2013 [16] a combination of those two practices TDD and CT, along with the continuous testing AutoTest.NET4CTDD open source plugin, was proposed and the preliminary feedback, via a survey inspired by Technology Acceptance Model (TAM), from developers in an industrial setting was collected. The new practice, combining CT and TDD was called Continuous Test-Driven Development or CTDD (the practice is described in Sect. 2). The findings of that paper were that CTDD could gain acceptance among TDD practitioners. Our initial speculations were that the benefits regarding development time could be small for a single developer, but could turn to be large due to the size of the software industry. There was no other scientific research published comparing TDD and CTDD. Hence, we decided to assert our initial speculations that the new practice could provide some time-related benefits in comparison to the usage of TDD. Our research goal and hypothesis are refined in Sects. 3.1 and 3.2, respectively.

In our previous paper [17], we argued that software engineering needs agile experimentation and formulated Agile Experimentation Manifesto with the aim to support lightweight, agile experimentation in industrial software development setting, where fully fledged experiments are often not feasible as it is not easy to involve professional software developers in a large-scale experimental research. Because of this natural limitation, in accordance with the first postulate of our Agile Experimentation Manifesto (Use small-n and single case experiments rather than large scale experiments to cut costs and enable experimentation), we decided to use the single-case/small-n experiment design (described in detail in Sect. 3.3). We use single developer to compare the TDD and CTDD practices in a baseline-intervention experiment design, where we will treat TDD as a baseline and CTDD as an intervention.

## 2  Background

As it was proposed by the authors [16], CTDD is a software development practice that combines TDD and Continuous Testing. If a developer uses TDD he needs to first write a test and verify if the tests fails (or the build fails because the functionality that is supposed to be tested was not written yet), then writes the functionality to quickly satisfy the test and executes it to check if it is the case, then he refactors the code regularly performing the tests to check if he did not break anything. We proposed to add the notion of continuous testing to TDD. The idea is that a developer that uses CTDD is not forced to perform the tests by himself. In continuous testing, the code is being compiled and tested automatically after the developer writes it and saves the file. So the need to manually trigger the testing was removed, potentially adding value to the process (via more frequent and earlier feedback from the amended code to the developer). In our earlier paper [16], we conducted a quick TAM inspired survey among the professional software developers that encouraged us to proceed with the new CTDD practice research (and supporting tool development), as it seems to be an improvement over the baseline TDD practice which could be widely accepted and adopted by practitioners. We make a conjecture that if the developers that use TDD embrace CTDD it can lead to small improvement in their coding performance that, taking into consideration a number of developers using TDD, could lead to noticeable savings within a company adopting the practice, not to mention the entire software industry.

AutoTest.NET4CTDD tool [16] that we co-developed makes it possible to use CTDD practice and can gather, in real time, various statistics during software development (the feature that the IDEs with built in CT, e.g., Visual Studio, are lacking). The tool was designed to execute the tests that are related to the changes the developer has made in the project. It detects what tests to run regardless of their type— unit, integration, system. As long as the tests are in the project and are related to the change they are run. If those tests are used as regression tests they will also be targeted by the AutoTest.NET4CTDD tool.

To allow empirical comparison of CTDD with the baseline TDD practice, we needed another tool for gathering the same data while using TDD. We did not found any suitable tool for the project setting we had access to. Hence, we had to develop our tool called NActivitySensor described in the appendix to our recent paper [17]. Having both tools, we were able to gather data needed to perform the empirical evaluation of the investigated software practices, TDD and CTDD. We were interested in the time that elapses from the moment when any of the tests fails, or the project does not build to the time where the project compiles, and all the tests are passing. We called this the red-to-green time (RTG time). We focused on the RTG time as it is where the advantage of the CTDD practice over the TDD practice may appear and be easily measured using the aforementioned tool set we developed.

# 3 Experiment Planning

## 3.1 Experiment Goal

The objective of the empirical study is to determine the differences in individual performance of a professional developer using CTDD versus TDD. Our quasi-experiment (as without random assignment of participants to different groups) is motivated by a need to understand the differences in individual performance while using the established TDD practice and the new CTDD practice introduced earlier by the authors [16].

The object of the study is the participant of a real software project. He is a professional software developer, computer science graduate with the MSc degree and 2 years experience (at the time of the experiment) in commercial software development.

The purpose of the quasi-experiment is to evaluate the individual performance when CTDD and TDD are employed. The experiment provides insight into what can be expected regarding individual performance when using the CTDD practice instead of TDD.

The perspective is from the point of view of the researchers, i.e., the researchers would like to know if there is any systematic difference in the CTDD versus TDD performance.

The main effect studied in the experiment is the individual performance measured by the RTG time introduced in Sect. 2. As mentioned earlier, the RTG time elapses from the moment the project is rendered not compiling or any of the tests is not passing, until the time the project builds, and all the tests are passing. The shorter the time, the more time the developer is spending doing actual work, i.e., development of new features. Because the task of executing the tests and waiting for their result is common and constantly recurring, we can assume that if we can reduce the time spent handling the tests we can reduce waste and make developers more productive.

The experiment is run within the context of the real software project in which a civil engineering software for calculation of concrete constructions was developed. The investigated developer developed modules for data exchange between this software and other data formats.

The summary of scoping of the study made according to the goal template by Basili et al. [3] is as follows: *Analyze* the CTDD practice *for the purpose of* evaluation *with respect to* its effectiveness *from the point of view of the* researcher *in the context of* a professional software developer in a real-world (industrial) software development project.

## 3.2 Hypothesis Formulation

An important aspect of experiments is to know and to formally state precisely what is going to be evaluated.

Null hypothesis, H0: There is no difference in the developer coding velocity, measured as the RTG time ($T_{RTG}$) introduced in Sect. 2, between the CTDD and TDD observation occasions, i.e., H0: $T_{RTG}(CTDD) = T_{RTG}(TDD)$

Alternative hypothesis, H1: $T_{RTG}(CTDD) < T_{RTG}(TDD)$

As CTDD reduces waste mentioned in Sect. 3.1, we may assume a directed hypothesis.

## 3.3 Experiment Design

We decided to perform the experiment in a real professional software development environment and on a real project to increase the external validity of the obtained results to the level hard to achieve with computer science students at a university lab. It would help us to generalize the results of our experiment to industrial practice. However, we need to take into account that the industrial resources willing to spend their precious time on research investigation instead of on commercial projects are scarce. Furthermore, we are aware that the goals of the different parties regarding experimentation are not necessary converging. For example, researchers are interested in performing controlled experiments in real projects resulting in reliable conclusions that lead to industry process improvement. The project owner is mainly interested in return of investment (ROI) and only secondarily in incorporating the results of experimentation provided they have a positive impact on the project itself. Furthermore, professional developers are expensive and busy people, while projects they are working on are seldom available for scientific research. Up front, we gave up the idea to perform a large scale experiment in an industrial setting, and we had no possibility to perform the project or even its part twice, once with the traditional (TDD) approach and once with the new one (CTDD), as it would require a lot of money. However, we have access to a small software development company based in Poland where one of the customers had a scientific background and was kind enough to allow some experimentation on a small staffed project. So we had a single object (professional software project) and single subject (developer) available for experimentation.

A good experimental design removes threats to internal validity, i.e., eliminate alternative explanations. A powerful tool in achieving internal validity is randomization. In classic large-$n$ experimental designs, treatments (interventions) are randomly assigned to subjects (participants). Unfortunately, it is unattainable in single-case studies, but it does not mean that it is impossible to randomly assign treatments to observation occasions. If we make a series of observations on a single case, or on a few cases, we can think of each as an observation occasion. As we randomly assign treatments to participants in large-$n$ experimental designs, we may randomly assign treatments to observation occasions in single-case or small-$n$ experimental designs. It is worth mentioning that classic parametric tests can not be used to analyze data coming from single-case or small-$n$ experimental designs, even if we do randomly allocate treatments to observation occasions. This is because parametric tests assume (apart from other assumptions) that the observations are independent, which is obviously not suited to the situation when we collect a series of measurements on a single case. Fortunately, there is a kind of tests, namely randomization tests, that fit great to the scenario. First of all, they do not require that the observations are independent. Secondly, they do not rely on the unrealistic assumption of random sampling from a population which is often not true in large-$n$ experiments. Concluding, single-case/small-$n$ experimental design combined with randomization tests provide us with a convenient design and analysis framework for agile experimentation in an industrial setting. Such research methods were used until now mainly in social psychology, medicine, education, rehabilitation, and social work [11]. Although they not entirely new in software engineering [10, 25].

We have access to a small project with two professional developers of which one will be working using TDD. Specific characteristics (constraints) related to our industrial software project environment are presented in Table 1.

The software project characteristics impose some constraints on the design of the experiment. Studying experimental designs discussed by Dugard et al. [8] we found a *single-case randomized blocks design with two conditions* to be the most appropriate because:

1. We have only one participant.
2. We have two conditions to compare: TDD (A) and CTDD (B).
3. We can arrange the two conditions in blocks.
4. It is possible to assign conditions to observation occasions in blocks at random—how we do that is described in Sect. 4.

We will have one participant (developer), and we expect minimum 40 classes (in 4 modules) to be relevant for our research. Having approximately 40 classes to observe

**Table 1** Software project under investigation

| No. of programmers | 2 |
| --- | --- |
| No. of programmers in experiment | 1 |
| No. of testers | 1 |
| Project time per month | 160 h |
| Project duration | 12 months |
| No. of modules | 37 |
| No. of TDD modules | Approx. 4–8 |
| No. of TDD classes | Approx. 40–80 |

development on and two treatments (TDD and CTDD) we would have 20 blocks[1] giving us $2^{20}$, i.e., a little over a million (1048576 to be exact), possible arrangements, which should give us good statistical power to detect difference between TDD and CTDD. We have used the Excel macro by Dugard et al. [8] to confirm our calculations.

## 4 Execution of Experiment

The experiment was conducted on an industry grade software project. The software that was being created is a construction engineering software for analysis and design of concrete constructions. Under investigation were all the classes from one module used specifically to calculate the units of measurement in the software. The customer from the United States for which the software was build agreed to perform experimentation on the project, but he issued one condition: the impact in terms of time and cost on the project should be minimal. It was agreed that the software engineers conducting the project would spend the minimal amount of time on the tasks not increasing the immediate ROI for the customer. The tools developed for this research proved to be sufficient for this task. The developers used the NActivitiSensor and AutoTest.NET4CTDD—the use of those tools does require no or minimal developer attention. The only manual thing required from developers to do was to use another tool for randomization. As mentioned in Sect. 3.3 we decided to use *single-case randomized blocks design with two conditions*. To introduce randomization, we implemented a small tool. This tool was used by the developers to randomly choose weather TDD or CTDD should be used for a given source code part. The tool is presented in Fig. 1. It randomly chooses weather the next class that a developer adds to the project should or should not to be decorated with a comment //AUTOTEST_IGNORE. If the comment is present the usage of AutoTest.NET4CTDD is disabled for this class, and the developer needs to execute the test manually. Classes without this comment are processed continuously tested. The act of "generating" the next random phase A (TDD) or B (CTDD) and possibly decoration the class with a comment is the only thing the developers needed to do to allow us to perform the experiment.

The data was gathered in the relational databases of NActivitiSensor and AutoTest.NET4CTDD. Post-processing of the data was done using a simple C# script that calculated the time from of the raw timestamped event data saved in the database. It calculated red-to-green (RTG) duration. RTG is the time in minutes that elapses either from the moment the project is not building to the moment the project is building properly, and all tests are passing or from the moment a test or tests are not passing to the moment when the project is building properly, and all tests are passing.

---

[1]There are always two possible arrangements for every block: first A then B or first B then A.

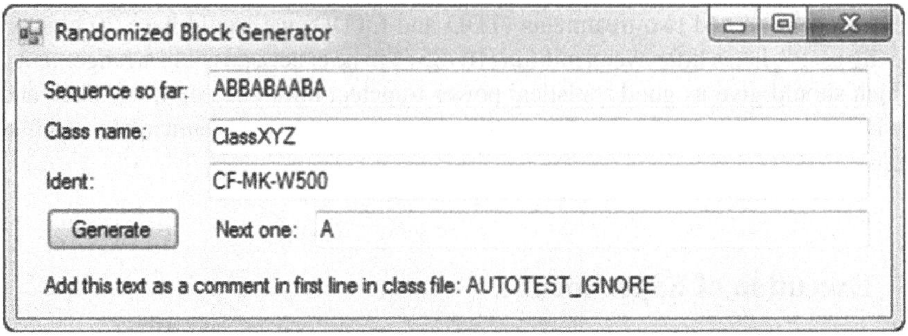

**Fig. 1** Random block generator tool

# 5 Analysis

For the statistical analysis of the data, we used the R environment with the SSDforR package. The former is a language and software environment for statistical computing [19], while the latter is the R package for analyzing single-subject data [2].

## 5.1 Descriptive Statistics

A typical way to begin comparing the baseline (TDD) and the intervention (CTDD) phases is by calculating descriptive statistics including measures of central tendency (e.g., mean, median, trimmed mean), as well as variation in both phases. It is often recommended to look for outliers and observations beyond two SDs are often checked, considered outliers and excluded from further analysis. We have followed this recommendation and found that the RTG times including the midnight (when the project with some failing test(s) was left until the next working day) were removed. Investigating further, we found that some developers want to know where to start in the next working day and that is why they leave a failing test.

Table 2 shows the descriptive statistics of the data collected during our empirical study. The RTG times are further visualized in Fig. 2, as well as using boxplot in Fig. 3 comparing the RTG time of the developer using TDD and CTDD. Visual examination of the boxplot indicates a slight drop of the RTG in the CTDD phase of the experiment.

The mean is a measure of central tendency helpful in describing the typical observation. Smaller RTG time mean in the intervention phase (CTDD) than in the baseline phase (TDD) suggests the positive impact of the CTDD practice ($M_{TDD} = 12.372$ vs. $M_{CTDD} = 8.388$). Unfortunately, the mean can be strongly influenced by outliers, especially when the number of observations is small.

The median, i.e., the value for which a half of the observations fall above and half below, is the only measure of central tendency that shows the advantage of TDD,

**Table 2** Descriptive statistics

| Measurement | A (TDD) | B (CTDD) |
|---|---|---|
| Number of observations | 85 | 55 |
| Median (Md) | 1.768 | 3.018 |
| Mean (M) | 12.372 | 8.388 |
| 10% trimmed mean (tM) | 7.713 | 5.67 |
| Standard deviation (SD) | 20.724 | 12.459 |
| Minimum | 0.275 | 0.295 |
| Maximum | 116.746 | 60.286 |
| IQR | 13.439 | 9.804 |
| 0% quantile | 0.275 | 0.295 |
| 25% quantile | 0.584 | 0.758 |
| 75% quantile | 14.023 | 10.562 |
| 100% quantile | 116.746 | 60.286 |

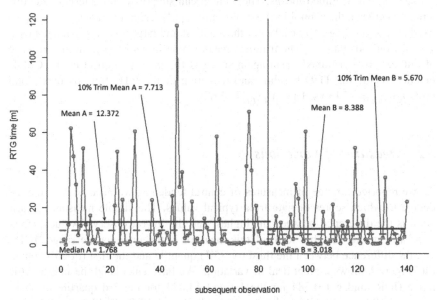

**Fig. 2** Subsequent RTG durations in phases A (TDD) and B (CTDD) with the mean and median imposed

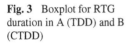

**Fig. 3** Boxplot for RTG duration in A (TDD) and B (CTDD)

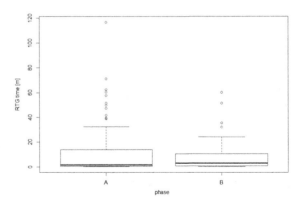

as suggests that the typical value of RTG time in TDD is smaller than in CTDD ($Md_{TDD} = 1.768$ vs. $Md_{CTDD} = 3.018$).

However, the recommended measure of central tendency in empirical software engineering are trimmed means. They which can support reliable tests of the differences between the central location of samples [12]. Trimming means removing a fixed proportion (e.g., 10 or 20%) of the smallest and largest values in the data set. The obvious advantage of the trimmed mean is that it can be used in the presence of outliers. 10% trimmed mean again suggests the positive impact of the CTDD practice versus the TDD baseline and reduction of the RTG time by over 2 min ($10\% \ tM_{TDD} = 7.713$ vs. $10\% \ tM_{CTDD} = 5.67$).

## 5.2 Measures of Variations

As we reported a range of measures of central tendency, we need to describe the degree to which scores deviate from typical values. The simple measure of variation is the difference between the highest and the lowest observed values. However, a bit more valuable form of this measure is the interquartile range (IQR), i.e., the difference between the third (or 75%) quantile and the first (or 25%) quantile. Figure 2 shows a great deal of variation. We have calculated the interquartile range (IQR) and got 0.584 for the 1st and 14.023 for the 3rd quartile in phase A, i.e., $IQR_{TDD} = 13.439$, and respectively 0.76 and 10.56 for the phase B, i.e., $IQR_{CTDD} = 9.804$. Hence, the variation in the middle 50% of the data is substantial.

The most common measure of variation frequently used together with the mean is the standard deviation (SD), which is the average distance between the scores and the mean. If the scores would be normally distributed then 95% of them would fall between 2 SDs below and above the mean, while typical scores fall between 1 SD below and above the mean. The standard deviation while using the CTDD practice is about half of the standard deviation of the baseline TDD practice ($SD_{TDD} = 20.724$

vs. $SD_{CTDD} = 12.459$) which is a desirable effect of CTDD. It is also easy to explain as the aim of the CTDD practice is to provide a fast feedback, to a developer, that tests do not pass.

## 5.3  Effect Size

Effect size is a name given to indicators that measure the magnitude of a treatment effect (CTDD vs. TDD in our case). Effect size measures are very useful, as they provide an objective measure of the importance of the experimental effect, regardless of the statistical significance of the test statistic. Also, effect sizes are much less affected by sample size than statistical significance and, as a result, are better indicators of practical significance [12, 15]. In the context of the performed empirical study, effect size quantifies the amount of change between the TDD and CTDD phases.

The SSD for R package calculates different kinds of effect size including $ES$ and $d - index$. ES is defined as the difference between the intervention (CTDD) and baseline (TDD) means divided by the standard deviation of the baseline. ES may show deterioration, lack of change or improvement due to the intervention.

$$ES = \frac{M_{intervention(CTDD)} - M_{baseline(TDD)}}{SD_{baseline(TDD)}} \tag{1}$$

In contrary to $ES$, $d - index$ uses a pooled standard deviation (i.e., a weighted average of standard deviations for two groups) which improves accuracy and is more appropriate when the variation between the phases differs, which is also the case in our study (see Table 2). However, it is worth mentioning that $d - index$ does not show the direction of the effect.

These effect size measures should only be used if there is no trend in the phases [13]. Each phase should be judged on the presence of the trend before continuing with the further investigation. If observations in any of the phase exhibit a trend, either decreasing or increasing, then measures of central tendency have limited abilities to correctly assess the response. In such case, neither mean nor median should be used [7], and the same applies to effect sizes $ES$ and $d - index$.

A trend may be defined and visualized by the slope of the best fitting line within the phase. The trends of both phases, A and B, were calculated using ordinary least squares (OLS) regression because it was found to be an accurate measure of the trend [7]. The multiple R-squared values were very close to 0 (0.005 for the phase A as well as B), while the p-values for the slopes in both phases were not statistically significant ($p > 0.05$), $p = 0.519$ for A and $p = 0.608$ for B. Hence, we may conclude that there were no (or were negligible) trends in the data.

As there was no trend in the data, we calculated effect size to measure the amount of change between A (TDD) and B (CTDD). The calculated effect sizes ($ES = -0.192$ and $d - index = 0.222$) can be interpreted as small, albeit non-zero, according to the guidelines provided by Bloom et al. [7].

## 6   Conclusions and Future Work

The results of our first quasi-experiment are rather inconclusive. The mean RTG time dropped in our experiment from 12.4 to 8.4 min, whereas the median slightly increased from 1.8 to 3 min. Both effect sizes ($ES = -0.192$ and $d - index = 0.222$) indicate a small degree of change (reduction) in the red-to-green time needed to satisfy the tests. The calculated effect sizes seem to suggest that there may be a slight impact of the application of the CTDD practice regarding the RTG time, which aligns with to some extent with our hypothesis. However, we need to remember that effect sizes alone, do not prove that the intervention was the cause of the observed change and further investigation is needed to obtain more reliable conclusions.

In our experiment the same developer uses TDD and CTDD which eliminates the threat of experience variability among the subjects but it raises the question whether the results of the experiment would scale across programmers of various backgrounds. That is one of the questions we would like to address in the follow up study we are planning on this topic.

From the informal interviews with the developers taking part in the quasi-experiment, we gathered positive feedback about the ease of use of the tool but noted slight dissatisfaction with the AutoTest.NET4CTDD tool overall performance. That might have an impact on the overall performance of the CTDD practice itself. In the short term, we will address those concerns by vertically scaling developer workplaces. To resolve the issue in the longer term, substantial work on the tool itself will be necessary.

Nevertheless, we were able to put our rules for Agile Experimentation Manifesto in motion while researching TDD versus CTDD. We noticed that we were able to incorporate state-of-the-art research techniques into a business-driven software project without affecting the project itself, which enable further experimentation. Using tools developed for this research we effectively minimized the time a developer need to spend on tasks not related to his core activities (e.g., related to research). Agile Experimentation applied in practice due to the course of the reported research appeared very promising to bridge the gap between research and industry in general, and developer, researcher and business owner in software engineering project in particular.

## References

1. Ambler, S.W.: How agile are you? 2010 Survey Results (2010). http://www.ambysoft.com/surveys/howAgileAreYou2010.html
2. Auerbach, C., Zeitlin, W.: SSDforR: Functions to Analyze Single System Data (2017). R package version 1.4.15
3. Basili, V.R., Caldiera, G., Rombach, H.D.: The goal question metric approach. In: Encyclopedia of Software Engineering. Wiley (1994)
4. Beck, K.: Extreme Programming Explained: Embrace Change. Addison-Wesley, Boston (1999)

5. Beck, K.: Test Driven Development: By Example. Addison-Wesley, Boston (2002)
6. Berłowski, J., Chruściel, P., Kasprzyk, M., Konaniec, I., Jureczko, M.: Highly automated agile testing process: an industrial case study. e-Inf. Softw. Eng. J. **10**(1), 69–87 (2016). doi:10.5277/e-Inf160104. http://www.e-informatyka.pl/attach/e-Informatica_-_Volume_10/eInformatica2016Art4.pdf
7. Bloom, M., Fischer, J., Orme, J.: Evaluating Practice: Guidelines for the Accountable Professional. Pearson/Allyn and Bacon (2008)
8. Dugard, P., File, P., Todman, J.: Single-case and Small-n Experimental Designs: A Practical Guide to Randomization Tests, 2nd edn. Routledge (2012)
9. Geracie, G.: The Study of Product Team Performance (2014). http://www.actuationconsulting.com/wp-content/uploads/studyofproductteamperformance_2014.pdf
10. Harrison, W.: N = 1: An alternative for software engineering research? (1997). Based upon an editorial of the same title in Volume 2, Number 1 of Empirical Software Engineering (1997). doi:10.1.1.5.2131&rep=rep1&type=pdf. http://citeseerx.ist.psu.edu/viewdoc/download
11. Kazdin, A.E.: Single-Case Research Designs: Methods for Clinical and Applied Settings. Oxford University Press (2011)
12. Kitchenham, B., Madeyski, L., Budgen, D., Keung, J., Brereton, P., Charters, S., Gibbs, S., Pohthong, A.: Robust Statistical Methods for Empirical Software Engineering. Empirical Softw. Eng. **22**(2), 579–630 (2017). doi:10.1007/s10664-016-9437-5
13. Kromrey, J.D., Foster-Johnson, L.: Determining the efficacy of intervention: the use of effect sizes for data analysis in single-subject research. J. Exp. Edu. **65**(1), 73–93 (1996). doi:10.1080/00220973.1996.9943464
14. Kurapati, N., Manyam, V., Petersen, K.: Agile software development practice adoption survey. In: Wohlin, C. (ed.) Agile Processes in Software Engineering and Extreme Programming. Lecture Notes in Business Information Processing, vol. 111, pp. 16–30. Springer, Berlin (2012)
15. Madeyski, L.: Test-Driven Development: An Empirical Evaluation of Agile Practice. Springer, Heidelberg (2010). doi:10.1007/978-3-642-04288-1
16. Madeyski, L., Kawalerowicz, M.: Continuous test-driven development - a novel agile software development practice and supporting tool. In: Maciaszek, L., Filipe, J. (eds.) ENASE 2013—Proceedings of the 8th International Conference on Evaluation of Novel Approaches to Software Engineering, pp. 260–267 (2013). doi:10.5220/0004587202600267. http://madeyski.e-informatyka.pl/download/Madeyski13ENASE.pdf
17. Madeyski, L., Kawalerowicz, M.: Software engineering needs agile experimentation: a new practice and supporting tool. in: software engineering: challenges and solutions. In: Advances in Intelligent Systems and Computing, vol. 504, pp. 149–162. Springer (2017). doi:10.1007/978-3-319-43606-7_11. http://madeyski.e-informatyka.pl/download/MadeyskiKawalerowicz17.pdf
18. Majchrzak, M., Stilger, Ł.: Experience report: introducing kanban into automotive software project. e-Inf. Softw. Eng. J. **11**(1), 41–59 (2017). doi:10.5277/e-Inf170102. http://www.e-informatyka.pl/attach/e-Informatica_-_Volume_11/eInformatica2017Art2.pdf
19. R Core Team: R: A Language and Environment for Statistical Computing. R Foundation for Statistical Computing, Vienna, Austria (2016)
20. Saff, D., Ernst, M.D.: Reducing wasted development time via continuous testing. In: Fourteenth International Symposium on Software Reliability Engineering, pp. 281–292. Denver, CO (2003)
21. Saff, D., Ernst, M.D.: An experimental evaluation of continuous testing during development. In: ISSTA 2004. Proceedings of the 2004 International Symposium on Software Testing and Analysis, pp. 76–85. MA, USA, Boston (2004)
22. Sochova, Z.: Agile Adoption Survey 2009 (2009). http://soch.cz/AgileSurvey.pdf
23. West, D., Grant, T.: Agile development: mainstream adoption has changed agility (2010). http://programmedevelopment.com/public/uploads/files/forrester_agile_development_mainstream_adoption_has_changed_agility.pdf

24. West, D., Hammond, J.S.: The forrester wave: agile development management tools, q2 2010 (2010). https://www.forrester.com/The+Forrester+Wave+Agile+Development+Management +Tools+Q2+2010/fulltext/-/E-RES48153
25. Zendler, A., Horn, E., Schwärtzel, H., Plödereder, E.: Demonstrating the usage of single-case designs in experimental software engineering. Inf. Softw. Technol. **43**(12), 681–691 (2001). doi:10.1016/S0950-5849(01)00177-X

# Enterprise Architecture Modifiability Analysis

Norbert Rudolf Busch and Andrzej Zalewski

**Abstract** As changes to organisations and the systems supporting them are getting ever more rapid, the modifiability of evaluation architecture is becoming ever more important. This paper presents a proposal of a scenario-based method for evaluating the modifiability of enterprise architectures. The method has been developed as an adaptation of the Software Architecture Analysis Method (SAAM) onto the field of enterprise architecture. It assumes that such architecture should be represented as a set of models in the ArchiMate language. The proposed method delivers an architecture assessment process together with techniques supporting the architecture assessment, such as representing requirements as scenarios of probable changes, analysing the impact of the changes represented by scenarios on enterprise architecture. The method has been validated on a real world example of an Enterprise Architecture of a Social Policy Support Entity of a certain municipality.

**Keywords** Enterprise architecture · Architecture assessment · Architecture evaluation

## 1  Introduction

The difference between enterprise architecture and software architecture is a matter of scale and the duration of lifetime. Enterprise architecture is rarely, if ever, replaced, meaning that it should be designed for long-term use. Software architecture can be relatively easily replaced, enterprise architecture not. Nevertheless, the surrounding world is constantly changing and requires modifications to architecture, so modifiability is now a key attribute for EA. At the same time, modifi-

N.R. Busch · A. Zalewski (✉)
Institute of Control and Computation Engineering, Warsaw University of Technology,
Warsaw, Poland
e-mail: A.Zalewski@ia.pw.edu.pl

N.R. Busch
e-mail: N.Busch@stud.elka.pw.edu.pl

© Springer International Publishing AG 2018
P. Kosiuczenko and L. Madeyski (eds.), *Towards a Synergistic Combination
of Research and Practice in Software Engineering*, Studies in Computational
Intelligence 733, DOI 10.1007/978-3-319-65208-5_9

ability is one of the quality attributes of IT systems, and the importance of the susceptibility of the enterprise architecture to changes is all the more essential as the systems evolve faster, which is a typical challenge that modern organisations face. The evaluation of architecture modifiability may deliver important insights into architecture changeability, which may facilitate the choice of more appropriate versions of architecture, or may reveal its flaws before commencing the development.

In order to evaluate the modifiability of enterprise architecture, a scenario-based evaluation method has been proposed. It was hypothesised that such a method could be defined by adapting one of the existing software architecture evaluation methods. In the end, the Software Architecture Analysis Method (SAAM) [1, 2] has been used for that purpose. The paper is organised as follows: related work is presented in Sect. 2, the method—in Sect. 3, an example validating the proposed method—in Sect. 4, the results are discussed in Sect. 5, Sect. 6 summarises the contribution and presents the research outlook.

## 2  Related Work

The term "enterprise architecture" is defined and discussed in a number of widely available studies, including standards, scientific and business papers and reports, as well as books, just to mention [3–7]. The enterprise architecture (architectural blocks) typically comprises (compare [6]):

- business architecture, namely, business goals and strategy, business processes and organisation's structure etc.;
- application architecture, which defines application systems, their interactions as well as their relationships to the business processes;
- technology architecture, which comprises technological components that constitute the platform for the operations of application elements;
- data architecture, i.e. enterprise data model, which defines the organisation's data assets.

The evaluation of enterprise architecture at early development stages enables its flaws to be detected and removed at minimal cost and time. Despite this, and the popularity and importance of the topic, the record of research in the field of enterprise architecture assessment is rather scarce.

The suitability of the Architecture Tradeoff Analysis Method (ATAM) [8] for an assessment of enterprise architecture has been analysed in workshops [9]. The outcomes of the analysis showed also that such an evaluation can be achieved by adapting one of the scenario-based software architecture analysis methods to the needs of enterprise architecture evaluation.

There is quite a broad knowledge about software architecture evaluation. A wide range of software architecture evaluation methods have been compared and analysed in [10–12]. As a modifiability assessment is the subject of this research, the natural reference is provided by three scenario-based methods, which enable or focus on architecture modifiability analyses, namely, SAAM [1, 2], ATAM [8] and the Architecture-Level Modifiability Analysis method (ALMA) [13].

In addition, much research has been conducted on enterprise architecture quality attributes analyses and its evaluation. As a result, the idea of EA quality attributes and their characterisation was proposed in [14], and several EA assessment methods and tools have been presented in [9, 15–22]. None of the above methods proposed a scenario-based approach focused on comprehensive enterprise architecture modifiability evaluation.

# 3 Enterprise Architecture Modifiability Analysis Method

The Enterprise Architecture Modifiability Analysis Method (EAMAM) is designed to measure the modifiability and adaptability of the architecture. The method has been designed as an adaptation of SAAM [1] a popular method for software analysis.

## 3.1 Evaluation Procedure

The method gives the organisation stakeholders the possibility to share a collection of probable scenarios of changes that the organisation may face in the future. These scenarios are then analysed and prioritised, before being mapped to the architecture views. This allows probable areas of higher complexity to be identified. Evaluations made using the EAMAM method require detailed documentation of the architecture, which allow for deep and comprehensive analysis.

Figure 1, in Business Process Model and Notation (BPMN), illustrates the procedure of evaluating an architecture using the EAMAM method. The evaluation process should be preceded by a review of an organisation with a presentation of the business goals of the enterprise architecture.

At the very outset of the evaluation using the EAMAM method, the available architectural description in the ArchiMate language is used, with the actions being based on this description. The analysis requires a significant amount of input data and consequently receives a lot of output data. This issue is described further later in this chapter.

### 3.1.1 P1 Scenario Identification

The first stage of the architecture evaluation is to determine its modifiability goals by using scenarios. The scenarios describe the needs of stakeholders in relation to

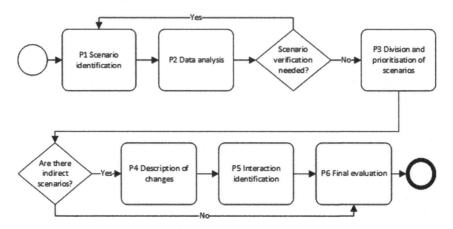

**Fig. 1** EAMAM method analysis procedure

the organisation. They explain the types of operations that the organisation must support, and also map all the changes that will be made within the organisation. When identifying scenarios, the organisation must firstly define their main representatives, anticipated changes and the properties by which the organisation should be identified, now and in the future. Scenarios play an analogical role as in software analysis methods [11].

Scenarios can be developed in several iterations. The identification of scenarios and data analysis should be performed alternately, as illustrated in Fig. 1. The increasing amount of data implies the identification of an expanding number of change scenarios.

**The Structure of Scenarios**

The scenario is a short statement describing the way in which the organisation intends to operate. It describes changes in a business, application or technology layer of an organisation, needed to accomplish a specific task. A single scenario may be relevant to many stakeholders.

### 3.1.2  P2 Data Analysis

The next step is the analysis of data. The architecture, or a number of potential architectures, must be represented in a way that is convenient for all the stakeholders. These representations are based on architectural views set out in the ArchiMate language. They contain components such as business, application, technology and data, as well as relationships between all these components. The purpose of data analysis is to share important information in terms of evaluation.

Data analysis is designed to verify scenarios that relate to the specific properties of the considered architecture. Scenarios, however, represent architectural requirements that are verified in the process of analysing the data.

### 3.1.3 P3 Division and Prioritisation of Scenarios

In this phase, the scenarios are grouped, divided into direct and indirect scenario groups, and then each scenario is given its priority. The purpose of grouping scenarios is to identify and generalise related scenarios. Direct scenarios are those that are already provided by the organisation's architecture and do not require any modifications to the architecture, while indirect scenarios are those that are not directly supported and require architectural changes.

Direct scenarios represent requirements that have already been addressed during earlier architectural design stages, and are similar to use cases in Unified Modeling Language (UML) notation, while the indirect scenarios describe the required changes to the way in which the organisation operates in the architecture. The execution of indirect scenarios requires modification to the architecture. They can be expressed by modifying existing elements of the architecture, adding new elements, adding relationships between existing elements, or by deleting an element or a relationship. Indirect scenarios serve to assess the extent to which the architecture is capable of developing in a direction that is relevant for the stakeholders. The division into direct and indirect scenarios was adapted from SAAM [1, 2].

In order to analyse in a limited time the conditions of the most important scenarios from the stakeholder's point of view, the scenarios need to be prioritised. Prioritisation takes place through voting. Each stakeholder receives an allocation of votes equal to the number representing 30% of all scenarios. For example, if there are 30 scenarios, each stakeholder will have nine votes. Stakeholders can freely distribute votes to each scenario. This prioritisation technique is described as cumulative voting [23].

When everyone has cast their votes, they are counted, and then all scenarios are organised in relation to the numbers of votes they received. Following that, a threshold is defined to determine all the scenarios that need to be analysed in detail. This can be a specific number of votes, if there is a clear limit (a clear gap in the number of votes cast), or it can be a point to be determined according to time limits.

### 3.1.4 P4 Description of Changes

The scenarios used for further evaluation are mapped as architectural representations. Each direct scenario should be presented to show how to implement it in the architecture. Each indirect scenario must include a description of the specific architectural changes required to implement it, as well as the estimated effort required for such implementations. Modifying the architecture means that either a new component or relation is added, or an existing component or a relation requires modification.

The description of all changes is recorded in the form of a table comparing all direct and indirect scenarios.

### 3.1.5   P5 Interaction Identification

The interaction between scenarios means that more than one scenario requires the modification of a single architecture element. The interaction identification explores how the operations of the organisation are separated. The interaction of scenarios that are not related show elements of architecture performing unrelated operations. These elements represent probable areas with higher complexity. In addition, the interaction study verifies whether the architecture representation was described at a sufficient level of structural decomposition.

The structural complexity of architectural components represents the number of interactions between scenarios. This number indicates those areas of architecture to be given special attention during further development work.

### 3.1.6   P6 Final Evaluation

During the final evaluation, the scenarios should be re-examined in terms of their relevance to the organisation's business objectives, taking into account the weightings assigned to scenarios during step P3 (Sect. 3.1.3). Indirect scenarios are assessed referring to the established criteria. Those may be the cost and time of implementing scenarios, or risks related to the implementation. Direct scenarios indicate that the stakeholder expectations are provided without the need to modify the architecture of the organisation.

In the case of comparing several potential architectures, an evaluation summary table should be drawn up. This table compiles scenarios with architectural proposals. Assigning weightings to each scenario unambiguously specifies the architecture that better meets the requirements described in the scenario. Architectural approaches are evaluated on the basis of intermediate scenarios, where a "+" indicates that the architecture requires fewer changes, a "−" that the architecture requires a significant number of changes, and a "0" where they both need the same changes. The ending result of architectural evaluation is received after giving the weightings to the scenarios and giving them the numerical symbols "+" and "−".

## 3.2   Evaluation Results

As a result of the analysis by the EAMAM method, a number of output data and a number of benefits are obtained. First and foremost, the enterprise architecture is reproducing planned changes to an organisation, which allows single elements that need to be modified to be identified, along with an indication of the amount of work related with implementing these modifications. The analytical result is obtained by assigning weightings to scenarios representing changes, and by defining the time necessary to implement those changes.

By evaluating several possible architectures, information is received about which of the architectures meet the requirements of scenarios with fewer modifications. In the event of a single architecture analysis, the information received covers the work efforts of implementing changes, and the complexity of architectural areas (architectural elements requiring modification). In addition, the evaluation allows the architectures to be analysed in terms of supporting business processes, before they are implemented.

# 4  Validation

The effectiveness of the EAMAM method was verified by applying it in practice to the analysis and evaluation of the enterprise architecture designed for a certain municipality. The case study described in this chapter provides the documentation of all the actions that were necessary to perform the EAMAM evaluation of the architecture and obtain the expected results.

## 4.1  Overview of the City Social Policy Enterprise Architecture

Prior to the implementation of enterprise architecture in the area of the social policy of the city and its units, the architectural approach was developed spontaneously. In particular subdivisions, there was no single coherent approach to the development of architecture, covering all business areas. Over the next few years, the management of architecture in both the business process layer and the business services, as well as in the technological aspects, was to be centralised.

The purpose of implementing enterprise architecture was to provide five main categories of support (application, technology and data) that were developed on the basis of overarching goals, business needs and development plans for the entities of the city's social policy:

1. Support in development of analytical and management tools
2. Support in data integration
3. Support in the development of electronic channels, keeping direct channels
4. Support in service-oriented integration of IT solutions
5. Support in striving for consolidation of IT solutions

Due to the strong influence of frequent legislative changes on the city's business processes and IT systems, the city office was expected to submit to the enterprise architecture of the City Social Policy an analysis of its adaptability and possible changes to its scope, which were planned based on the five support categories mentioned above. The architecture attribute that was to be examined was its modifiability. In order to verify the suitability of the architecture for achieving the objectives, it was decided to evaluate the architecture using the EAMAM method. The following subsections describe the effects of completing each step of the procedure. The two candidate architectures of the City Social Policy organisation were evaluated.

## 4.2   P1 Scenario Identification, First Iteration

Scenarios for the City Social Policy organisation were developed in two iterations. Since the architecture had not yet been analysed, the first iteration of identifying scenarios was carried out in the form of a series of proposals based solely on the requirements for the organisation. As a result of the first iteration, a list of unprocessed scenarios was created.

## 4.3   P2 Data Analysis, First Iteration

The enterprise architecture views of the City Social Policy were designed, including all eighteen standard architectural viewpoints in the ArchiMate language. All the viewpoints were important for the overall design, but only some of them contained information related to this specific EAMAM evaluation. The architecture views representing organisation and application co-operation viewpoints presented appropriate data and were further analysed.

## 4.4   P1 Scenario Identification, Second Iteration

Stakeholders began the second iteration of the scenario identification process, based on the knowledge raised in the architectural analysis step. The result was 13 new scenarios.

## 4.5 P2 Data Analysis, Second Iteration

Two candidate architectures for IT systems were then analysed. The main goal of the evaluation was to decide which architecture for the information systems to adapt, and the EAMAM method provided a tool enabling the analysis outcomes to be based on analytical foundations. The first architecture was based on a consolidation and centralisation model, while the other proposed to remain on a distributed model.

Figure 2 shows the view of applying co-operation of the first candidate architecture, which is based on a centralised model. The centralised version assumes the implementation of one Central City Social Policy System, providing a full service in a uniform, coherent system. It is assumed that in this variant, the Central City Social Policy System will have to take over the functionality currently supported by domain systems. It is also planned to implement a central financial and accounting system, a central HR system, and document management system for the city hall, with the possibility of implementing these solutions in subordinate units. It is also assumed to establish a single and comprehensive service catalogue in the area of City Social Policy. Its creation is aimed at facilitating the monitoring of the effectiveness of services, i.e. all the services/forms of assistance offered to clients, while on the other hand also helping customers navigate the vast array of help available to them. The service catalogue will be one central information repository of services within the City of Social Policy. The ESB service bus of the Central Information Exchange Platform integration platform will provide information

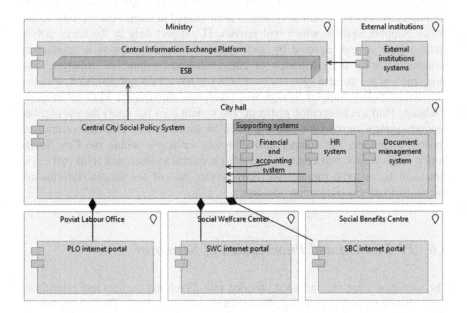

**Fig. 2** Application co-operation view of the architecture based on the centralised model

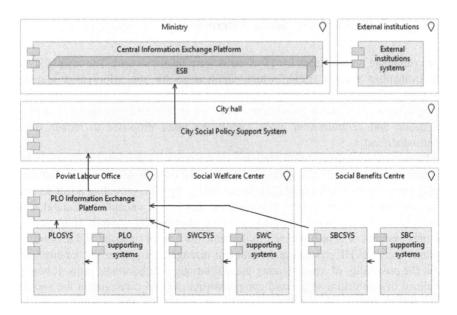

**Fig. 3** Application co-operation view of the architecture based on the distributed model

exchange between central and domain systems affecting the smooth operation of the social policy area, and will enable communication with third party systems.

Figure 3 shows the view of application co-operation of the second architecture, which was based on a distributed architectural model. This option does not include the implementation of a central system. It is envisaged to implement the City Social Policy Support System, which will provide IT support only in the social policy areas, which are not covered by such support, or where support is insufficient (without ensuring the integration of client information into one solution). The integration of the domain systems, i.e. PLOSYS, SWCSYS and SBCSYS, with the City Social Policy Support System will take place through the PLO Information Exchange Platform integration platform. The variant does not imply the use of other support systems beyond the existing ones. In addition, just as in the centralised version, a uniform and comprehensive service catalogue within the City Social Policy is envisaged, and communication with central systems and third party systems will be ensured through the ESB service bus of the Central Information Exchange Platform integration platform.

## 4.6 P3 Division and Prioritisation of Scenarios

The scenarios were then grouped, divided into direct and indirect and prioritised. After carrying out the grouping, 25 scenarios were received. Then the scenarios were prioritised. Every voter was given a number of votes equal to 30% of the

**Table 1** List of scenarios after prioritisation

| Scenario number | Number of votes | Scenario description |
|---|---|---|
| 4 | 9 | Possibility of adding e-services in the future, including electronic requests |
| 6 | 7 | Possibility of accessing data in real time, including data exchange in that mode with connected domain systems |
| 8 | 7 | Connecting the system with the client's portal, giving the client, after the authentication within the client's account, the possibility to fully view his data collected within the system |
| 9 | 6 | Examination of the degree of effectiveness in the common selection of services based on an analysis of support effectiveness, in various configurations, using analytical tools |
| 16 | 5 | Advising/running the application process, by verification of the completeness/correctness of incoming data, advising the employee what should be verified/what to pay attention to while providing the customer service |
| 20 | 5 | Giving the employee the possibility to view a fully examined case (preserving the rights to access suitable data related to the areas of the performed duties/purpose for processing data), including the verification of data necessary to make a decision in relation to granting a certain service |
| 22 | 4 | Support for document management, including the possibility of setting business rules of passing cases and supporting the passing of cases between domain systems |
| 25 | 3 | Alignment and automation of data update regarding the clients— support for management in terms of finding and deleting duplicates, standardisation, automatic and mass rules of validation usage |

number of scenarios. This meant that seven votes were to be allocated by each stakeholder. The eight scenarios with highest score were further examined.

The list of scenarios selected after prioritisation is presented in Table 1. The scenarios listed in Table 1 were then divided, with reference to both architectures, to direct and indirect. Only one scenario turned out to be direct, and only for the architecture based on the distributed model. The other scenarios were all indirect for both architectures.

## 4.7 P4 Description of Changes

After the division of scenarios, the estimated effort required to execute each scenario was analysed. The outcome was presented in two tables—one for the architecture based on the centralised model, and the second one for the architecture based on the distributed model. Table 2 shows a fragment of the table representing scenarios for architecture based on the centralised model.

**Table 2** Fragment of the evaluation of scenarios for the architecture based on the centralised model

| Scenario number | Scenario description | Category | Modified components | Estimated effort |
|---|---|---|---|---|
| 4 | Possibility of adding e-services in the future, including electronic requests | Indirect | Central city social policy system | 9 man-months |
| 6 | Possibility of accessing data in real time, including data exchange in that mode with connected domain systems | Indirect | Central city social policy system | 3 man-months |
| 8 | Connecting the system with the client's portal, giving the client, after the authentication within the client's account, the possibility to fully view his data collected within the system | Indirect | Client's panel added | 12 man-months |

## 4.8 P5 Interaction Identification

In addition to the above estimates, architectural areas affected by several distinct scenarios were identified, by mapping each scenario from Table 1 to the architectural views. As a result, areas of scenario interactions were indicated.

Figure 4 illustrates scenarios mapped to the distributed architecture view. Two sets of linked scenarios were formed. The first one included scenarios 6 and 9, and

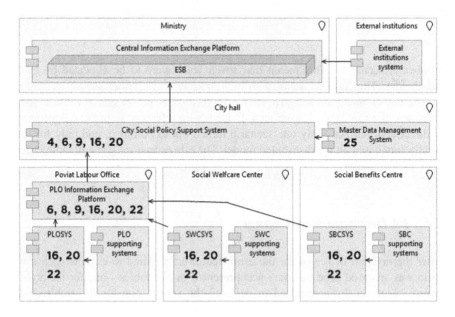

**Fig. 4** Scenarios mapped to the application co-operation view of the architecture based on the distributed model

the second one included scenarios 16, 20 and 22. It was identified that both sets of scenarios operated within City Social Policy Support System and the PLO Information Exchange Platform integration platform.

## 4.9   P6 Final Evaluation

The EAMAM method evaluation result is shown in Table 3. The centralised architecture proved able to support the analysed scenarios better than the architecture based on the distributed model, when taking into account the scenarios with highest weightings. Furthermore, the evaluation process implied other changes to the architecture. It was decided to implement a Data Warehouse, including a Business Intelligence (BI) module. This tool, using dedicated Extract, Transform, Load (ETL) connectors, was to collect information from the Central City Social Policy System on the social policy actions, their costs and effects. To integrate information about the client and to unify access to specific thematic areas, both to the labour market and to social security, it was decided to launch the Client Portal under the subpage of the current city hall website. In order to support the transfer of cases between individuals and to provide a complete view of the case, it was decided to implement a "Workflow" system. The extension of city hall's infoline services was assumed by adding an additional service relating to the area of social policy—infoline staff will provide the same information that is available on the Client Portal, for those who cannot or do not know how to use the internet.

## 5   Discussion, Limitations

As a result of this research, a method for analysing and evaluating the modifiability of enterprise architectures represented in the ArchiMate language was developed. The hypothesis set out at the beginning of the studies, namely that this method can be developed by adapting existing methods for software architecture analysing, has been confirmed by the successful adaptation of the commonly used method— SAAM.

**Table 3** Final evaluation summary

| Scenario number | Centralised model | Distributed model |
|---|---|---|
| 4 | 0 | 0 |
| 6 | + | − |
| 8 | + | − |
| 9 | + | − |
| 16 | + | − |
| 20 | + | − |
| 22 | − | + |
| 25 | + | − |

The EAMAM method is primarily intended to support in responding to the requests that rise during the development of the organisation, which require changes and assessments of the architectural decisions, which, as a result, lead to the fulfilment of stakeholders' requirements.

There are a number of scenario-based methods for software architecture analysis that could potentially apply in the field of enterprise architecture modifiability evaluation, i.e. SAAM, ATAM, ALMA, but each of these methods assumes a description of the architecture during the evaluation process. In the case of enterprise architecture and its scale, it is infeasible to describe the architecture during the evaluation process. Furthermore, enterprise architecture is usually represented in several languages, e.g. BPMN for business layer views, UML for application layer views, and ER models for data layer views. To enable a comprehensive evaluation of the modifiability of enterprise architectures allowing the identification of implications caused by changes to components of one architectural layer on elements of other layers, the necessary prerequisite for the evaluation process is the description of the architecture in a unified language. The method of analysis and evaluation described in this paper imposes the ArchiMate language for a coherent architectural description and further evaluation.

In addition, SAAM (in its final shape [24]) does not focus on a specific quality attribute. The scenarios raised by the stakeholders determine which system qualities are investigated. The emphasis of ATAM is also not on an in-depth assessment of a single quality attribute, but rather on identifying trade-offs between quality attributes. ALMA does require some pre-development (pre-architecting) of changes when analysing the impact of changes. This may be unachievable in the case of enterprise architecture, as architecting is usually an even more challenging and time consuming than evaluation.

The contribution of this paper is to prove that the classic method of architecture analysis—SAAM—with its technique of classifying scenarios into direct and indirect classes, works also well in the case of enterprise architectures. Architectural models in the ArchiMate language should be provided as an input, instead of the architecture description step during the evaluation process, but all the other SAAM steps fit into the enterprise architecture evaluation.

## 6  Summary and Research Outlook

The method described in this paper is a tool for evaluating enterprise architectures. It was developed for the purpose of analysing and evaluating in advance the requirements for architecture regarding modifiability. Replacing the requirements with scenarios that specify the requirements, the method achieves its purpose. By defining scenarios, it is possible to shape the most important business goals, and to explore the mapping of these scenarios to the analysed architectures.

When asked if it is possible to systematically analyse and evaluate enterprise architectures to help quickly identify problems in the organisation's design and to reduce the cost of mistakes, the answer must be—yes, at least in the case of modifiability.

As the global rivalry of organisations is growing increasingly intense every day, and the pace of development of enterprise architectures is growing faster and faster, one of the key challenges is to develop general purpose methods used to analyse several quality attributes, as well as special purpose methods used to evaluate specific organisational models. In addition, it is worth examining the characteristics and differences of enterprise architectures used in various economy sectors. A challenge may also be to build a tool that automates the evaluation of enterprise architectures, based on entered an architectural model's requirements for quality attributes.

# References

1. Kazman, R., Webb, M., Abow, G., Bass, L.: SAAM: a method for analyzing the properties of software architectures. In: Proceedings of 16th International Conference on Software Engineering, pp. 81–90 (1994)
2. Kazman, R., Clements, P., Abowd, G., Bass, L.: Scenario-based analysis of software architecture. IEEE Softw. 13(6), 47–55 (1996)
3. Zachman, J.: A framework for information systems architecture. IBM Syst. J. 26(3), 76–292 (1987)
4. Lankhorst, M.: Enterprise Architecture at Work: Modelling, Communication, and Analysis. Springer, Berlin (2013)
5. Bernard, S.A.: An Introduction to Enterprise Architecture. AuthorHouse, Bloomington (2012)
6. The Open Group, *TOGAF Version 9.1*, Van Haren Publishing, Zaltbommel, the Netherlands (2011)
7. Weill, P., Robertson, D.C., Ross. J.W.: Enterprise Architecture as Strategy: Creating a Foundation for Business Execution. Harvard Business Review Press, Boston (2006)
8. Klein, M., Clements, P., Kazman, R.: ATAM: Method for Architecture Evaluation. In: CMU/SEI-2000-TR-004, Software Engineering Institute, Pittsburgh, USA (2000)
9. Klein, J., Gagliardi, M.: A Workshop on Analysis and Evaluation of Enterprise Architectures, CMU/SEI-2010-TN-023. Software Engineering Institute, Pittsburgh, USA (2010)
10. Zalewski, A.: Modelling and evaluation of software architectures. Prace Naukowe Politechniki Warszawskiej. Elektronika 187, 3–116 (2013)
11. Kazman, R., Klein, M., Clements, P.: Evaluating Software Architectures: Methods and Case Studies. Addison-Wesley, Boston (2002)
12. Dobrica, L., Niemelä, E.: A survey on software architecture analysis methods. IEEE Trans. Softw. Eng. 28(7), 638–653 (2002)
13. Bengtssona, P., Lassing, N., Bosch, J., Vliet, H.: Architecture-level modifiability analysis (ALMA). J. Syst. Softw. 69, 129–147 (2004)
14. Davoudi, M.R., Aliee, F.S.: Characterization of enterprise architecture quality attributes. In: 2009 13th Enterprise Distributed Object Computing Conference Workshops, pp. 131–137 (2009)
15. Ekstedt, M., Franke, U., Johnson, P., Lagerström, R., Sommestad, T., Ullberg, J., Buschle, M.: A tool for enterprise architecture analysis of maintainability. In: 2009 13th European Conference on Software Maintenance and Reengineering, pp. 327–328 (2009)

16. Johnson, P., Johansson, E., Sommestad, T., Ullberg, J.: A tool for enterprise architecture analysis. In: 11th IEEE International Enterprise Distributed Object Computing Conference (EDOC 2007), p. 142 (2007)

17. Javanbakht, M., Pourkamali, M., Derakhshi, M.F.: A new method for enterprise architecture assessment and decision-making about improvement or redesign. In: 2009 Fourth International Multi-Conference on Computing in the Global Information Technology, pp. 69–78 (2009)

18. Javanbakht, M., Rezaie, R., Shams, F.: A new method for decision making and planning in enterprises. In: 2008 3rd International Conference on Information and Communication Technologies: From Theory to Applications, pp. 1–5 (2008)

19. Lakhrouit, J., Baïna, K.: Evaluating enterprise architecture complexity using fuzzy AHP approach: application to university information system. In: 2015 IEEE/ACS 12th International Conference of Computer Systems and Applications (AICCSA), pp. 1–7 (2015)

20. Davoudi, M.R., Aliee, F.S.: A new AHP-based approach towards Enterprise Architecture quality attribute analysis. In: 2009 Third International Conference on Research Challenges in Information Science, pp. 333–342 (2009)

21. Bijarchian, A., Ali, R.: A model to assess the usability of enterprise architecture frameworks. In: 2013 International Conference on Informatics and Creative Multimedia, pp. 33–37 (2009)

22. Song, H., Song, Y.: Enterprise architecture institutionalization and assessment. In: 2010 IEEE/ACIS 9th International Conference on Computer and Information Science, pp. 870–875 (2010)

23. Moore, B.F.: The history of cumulative voting and minority representation in Illinois, 1870–1908. Univ. Ill. Bull. **23** (1909)

24. Kazman, R., Bass, L., Clements, P.: Software Architecture in Practise. Addison-Wesley, Boston (1997)

# A Survey Investigating the Influence of Business Analysis Techniques on Software Quality Characteristics

Katarzyna Mossakowska and Aleksander Jarzębowicz

**Abstract** Business analysis is recognized as one of the most important areas determining the outcome (success or failure) of a software project. In this paper we explore this subject further by investigating the potential impact of techniques applied in business analysis on essential software quality characteristics. We conducted a literature search for software quality models, analyzed the existing models and selected a subset of commonly recognized quality characteristics. Also, we identified a representative set of recommended state-of-the-art business analysis techniques. These two sets provided the basis for questionnaire survey and interviews. We conducted a survey involving 20 industry professionals, followed up by 2 interviews with experienced business analysts to discuss and interpret survey results. The main outcome are recommendations regarding techniques to be used in software project for a given quality characteristic considered essential.

**Keywords** Business analysis · Requirements engineering · Software quality · Techniques · Survey

## 1 Introduction

The success of a software project is still uncertain, as numerous reviews of past projects indicate that a significant percentage of them ends up as failed or challenged by various problems [7, 29]. Traditionally, project success is defined in terms of time, budget and result, where the result, apart from developed product's scope, also includes product's quality. Alternatively, quality can be distinguished as a separate success criterion (e.g. [22]). However, whether quality is explicitly ref-

K. Mossakowska · A. Jarzębowicz (✉)
Department of Software Engineering, Faculty of Electronics, Telecommunications and Informatics, Gdańsk University of Technology, Gdańsk, Poland
e-mail: olek@eti.pg.gda.pl

K. Mossakowska
e-mail: katarzyna.mossakowska@gmail.com

© Springer International Publishing AG 2018
P. Kosiuczenko and L. Madeyski (eds.), *Towards a Synergistic Combination of Research and Practice in Software Engineering*, Studies in Computational Intelligence 733, DOI 10.1007/978-3-319-65208-5_10

erenced or not, it is commonly understood that it essential term in software engineering and project management and an important criterion when evaluating project's outcome.

Causal analyses of project failures and problems reveal that many commonly occurring contributing factors can be mapped to requirements engineering/business analysis (RE/BA) activities and practices [2, 3, 29]. In our research we tried to explore this subject further, in particular with respect to the influence of RE/BA on the quality of the developed system. As RE/BA is a complex domain in software engineering and includes many components (processes, techniques, competencies, good practices, software supporting tools etc.), we decided to narrow our research down to RE/BA techniques. Such techniques are described in many sources e.g. [12, 23, 31] as tools to be used by business analysts for e.g. requirements elicitation. We defined the following research question: *Which RE/BA techniques applied in a software project have the greatest influence on particular quality characteristics of the resulting software system?*

To answer it, first we conducted a literature search for software quality models, analyzed the existing models and selected a subset of commonly recognized quality characteristics. Also, we identified a representative set of recommended RE/BA techniques. These two sets provided the basis for questionnaire survey and interviews. We conducted a survey involving 20 industry professionals, followed up by 2 interviews with experienced business analysts to discuss and interpret results.

The remainder of this paper is structured as follows. In Sect. 2 we outline the related work. Section 3 describes the initial steps of our research—identifying RE/BA techniques (Sect. 3.1) and quality characteristics (Sect. 3.2). The main steps of conducted survey are described in Sect. 4, including: questionnaire design, data gathering and processing, interviews to interpret results and final results. We end this paper with validity threats discussion (Sect. 5) and conclusions (Sect. 6).

## 2   Related Work

There is a substantial amount of work published on evaluating techniques used in requirements engineering (RE) or more generally in business analysis (BA). Some researchers focused on techniques for a particular activity e.g. requirement elicitation [30], specification [4] or validation [16], others included a broader spectrum of techniques [15, 17, 18]. RE/BA techniques were evaluated with respect to: their inherent characteristics and potential [4, 15, 17, 30], applicability context (project size, product type etc.) [15, 17] and ability to address RE problems [16, 18].

Several studies exploring the influence of RE/BA process on developed product, project results or even more general outcomes were conducted [6, 10, 26, 27]. Hofmann and Lehner [10] identified a set of best RE practices leading to project's success. Sommerville and Ransom [27] reported that improvements to RE process maturity led to business benefits in all 9 companies participating in the study, however it was not possible to establish a strong causal link. A survey including over 400 companies by Ellis and Berry [6] revealed that higher level of company's maturity in RE and management processes correlates to the better success ratio of projects developed by such company. Sethia and Pillai [26] provided an analysis (based on a systematic literature review) of the negative impact of requirements elicitation problems on software quality and project's outcome. Also, a need and opportunity for further research providing a better understanding of RE effects and influences is noted [8, 11].

No work directly addressing impact of RE techniques on quality of the developed system could be found, except [24, 25], which uses a completely different research method—data mining for a large set of software projects' data.

## 3　Literature Search and Analysis

The first part of our research aimed at identifying valid input to be used in the survey. It was based on literature search and analysis of identified sources. We chose to combine information from several sources, instead of relying on a single source. We also had to make decisions to focus on the most essential items and leave out the others (to ensure that the survey is realistic with respect to its scope and number of questions asked). Two main areas of background were important for our research: RE/BA techniques and quality models which translate the generic "quality" term into more detailed characteristics and attributes.

### 3.1　Identification of BA Techniques

We decided to use two main sources to identify and select RE/BA techniques for the planned survey. First of them was an industrial standard: Business Analysis Body of Knowledge (BABOK) version 3 [12]. It is considered a renowned source for business analysts and a basis for CBAP (Certified Business Analysis Professional) certification process. Moreover, this most recent 3rd version had been published only several months before we started this research, so we considered it a state-of-the-art resource. As a second source we selected a book by Wiegers and Beatty [31], a comprehensible guidance covering a broad spectrum of software requirements topics. Its 3rd edition, published in 2013, was expanded with new themes e.g. requirements in agile development.

We analyzed both sources to identify techniques they recommend. As a next step, we selected a subset of them to keep the survey scope realistic. This led us e.g. to reject various kinds of diagrams and notations used to specify and document requirements, as there were too many of them to include them all and we wanted to avoid arbitrary selection. We intended to cover all areas related to RE/BA, not to e.g. restrict the survey to specification techniques only.

**Table 1** Selected RE/BA techniques and their sources

| Technique | Source(s) |
|---|---|
| *Area: elicitation* | |
| Scope modeling | [12] 10.41; [31], Chap. 5 |
| Stakeholder list, map or personas | [12] 10.43; [31], Chap. 6 |
| Focus groups | [12] 10.21; [31], Chap. 7 |
| Brainstorming | [12] 10.5 |
| Event-response lists | [31], Chap. 12 |
| Interviews | [12] 10.25; [31], Chap. 7 |
| Survey/questionnaire | [12] 10.45; [31], Chap. 7 |
| Document analysis | [12] 10.18; [31], Chap. 7 |
| Observation | [12] 10.31; [31], Chap. 7 |
| *Area: analysis* | |
| Organizational modeling | [12] 10.32 |
| Prototyping | [12] 10.36; [31], Chap. 15 |
| Prioritization | [12] 10.33; [31], Chap. 16 |
| Data dictionary | [12] 10.12; [31], Chap. 13 |
| Business model canvas | [12] 10.8 |
| SWOT analysis | [12] 10.46 |
| Risk analysis and management | [12] 10.38; [31], Chap. 32 |
| *Area: specification* | |
| SRS templates | [31], Chap. 10 |
| Item tracking | [12] 10.26 |
| Non-functional requirements analysis | [12] 10.30; [31], Chap. 14 |
| *Area: validation* | |
| Reviews | [12] 10.37; [31], Chap. 17 |
| Retrospectives | [31], Chap. 17 |
| Test cases and scenarios | [31], Chap. 17 |
| Acceptance and evaluation criteria | [12] 10.1; [31], Chap. 17 |
| *Area: management* | |
| RE planning | [31], Chap. 7 |
| Estimation | [12] 10.19; [31], Chap. 19 |
| Trainings | [31], Chap. 4 |
| Glossary | [12] 10.23; [31], Chap. 13 |
| Functional decomposition | [12] 10.22 |
| Roles and permissions matrix | [12] 10.39; [31], Chap. 2 |
| Lessons learned | [12] 10.27; [31], Chap. 31 |
| Metrics and key performance indicators (KPI) | [12] 10.28 |

The selected techniques are shown in Table 1, grouped by areas of application. The abovementioned sources introduce different classifications: [31] uses "classical" RE-related areas (Elicitation, Analysis, Specification, Validation, Management), while [12] defines 6 Knowledge Areas (BA Planning and Monitoring,

Elicitation and Collaboration, Requirements Life Cycle Management, Strategy Analysis, Requirements Analysis and Design Definition, Solution Evaluation). As these two classifications are not "compatible", we decided to use classification from [31] and do the "mapping" for the remaining techniques.

Techniques in Table 1 are also provided with references to the exact sections/chapters, where a given technique is described in sources(s). Most of these techniques appeared in both sources, however some techniques recommended by a only one of them were also selected because of their potential influence on quality.

## 3.2 Identification of Quality Characteristics

Since late 70s, several attempts to develop a generic software quality model were made e.g. [1, 5, 9, 13, 14, 19]. Basically, all models have a similar construction—they define a number of main quality characteristics (or attributes, factors etc. as different names are used), which are in turn decomposed into more detailed sub-characteristics. Such decomposition continues (number of its steps also varies between models) until measurements are possible and metrics can be defined.

We compared a number of generic quality models to identify the most common quality characteristics and establish their definitions. This task may appear unnecessary, as software quality models have already been analyzed, improved and compared (e.g. according to inclusion of particular characteristics [28] or deficiencies [20]). Our purpose was however different—instead of identifying the most complete or suitable model, we intended to identify a small number of most common quality characteristics included in established quality models and to define them by compiling proposals from several models. To achieve it, we conducted the analysis of 5 quality models, independent from existing published comparisons. The following models were analyzed: McCall model [19], Boehm model [1], Dromey model [5], FURPS model [9] and ISO 9126 model [13].

A comparison of quality characteristics included in those models is presented in Table 2. As different names and hierarchical decompositions are used in analyzed models, contents of Table 2 are the result of several decisions e.g.:

- The characteristics are on the same level of abstraction (e.g. Resource utilization is a part of Efficiency), but nevertheless we included them to provide a more comprehensive comparison.
- When alternative names to the same characteristic were given in various models, we tried to choose the one more consistent with current terminology (e.g. used in ISO 25010, which was not part of this comparison, but an additional reference in case of such conflicts).
- Only a more general Usability characteristic was included with assumption that it covers characteristics like As-Is-Utility and Human Engineering from Boehm model, as well as Usability sub-attributes from ISO 9126.

**Table 2** Comparison of characteristics in software quality models

|     | Characteristic/attribute/factor | Inclusion in quality models | | | | |
| --- | --- | --- | --- | --- | --- | --- |
|     |                                 | McCall | Boehm | Dromey | FURPS | ISO 9126 |
| 1.  | Accessibility        |   | x |   |   |   |
| 2.  | Adaptability         |   |   |   |   | x |
| 3.  | Analysability        |   |   |   |   | x |
| 4.  | Co-existence         |   |   |   |   | x |
| 5.  | Completeness         | x | x |   |   |   |
| 6.  | Correctness          | x |   | x |   |   |
| 7.  | Efficiency           | x | x | x |   | x |
| 8.  | Flexibility          | x |   |   |   |   |
| 9.  | Functionality        |   |   | x | x | x |
| 10. | Installability       |   |   |   |   | x |
| 11. | Integrity            | x | x |   |   |   |
| 12. | Interoperability     | x |   |   |   | x |
| 13. | Learnability         |   |   |   |   | x |
| 14. | Maintainability      | x | x | x | x | x |
| 15. | Modifiability        |   | x |   |   | x |
| 16. | Operability          | x |   |   |   | x |
| 17. | Performance          |   |   |   | x |   |
| 18. | Portability          | x | x | x |   | x |
| 19. | Reliability          | x | x | x | x | x |
| 20. | Replaceability       |   |   |   |   | x |
| 21. | Resource utilization |   |   |   |   | x |
| 22. | Reusability          | x |   | x |   |   |
| 23. | Suitability          |   |   |   |   | x |
| 24. | Supportability       |   |   |   | x |   |
| 25. | Testability          | x | x |   |   | x |
| 26. | Understandability    |   | x |   |   | x |
| 27. | Usability            | x | x | x | x | x |

The comparison provided the basis for selection of quality characteristics to be used in the planned survey. We intended to use a small number of more general, but well-defined characteristics. Table 2 reveals that 3 characteristics (Maintainability, Reliability and Usability) are explicitly addressed in all of analyzed models. Functionality is not included in older models (McCall, Boehm), but its importance is obvious. Therefore, we finally selected 4 characteristics, and (combining proposals from various models) defined them. The model we mostly relied on turned out to be ISO 9126, but the resulting characteristics differ from the ones in ISO 9126, because of influences from other models:

- Functionality—suitability to provide an appropriate set of functions, resulting with the needed degree of precision, ability to interact with specified external systems, assurance to prevent unauthorized users/systems access.
- Usability—capability to be understood, learned and operated by users, as well as being recognized by them as attractive.
- Reliability—capability to tolerate faults, to avoid failures and to recover in case of failure.
- Maintainability—capability to be modified and to be adopted to another environment with adequate performance.

# 4 Questionnaire Survey

In this section we describe the key elements related to the survey and its results.

## 4.1 Questionnaire Design

We planned to answer question about RE/BA techniques' impact on quality by conducting a questionnaire-based survey published in the Internet. A number of online survey software tools (SurveyMonkey, Google Forms, Interankiety, SurveyGizmo) were considered. The final choice was Google Forms—this tool lacked questionnaire design flexibility and user interface configurability, but provided the best functionality for reviewing and processing answers and was freely available without any restrictions.

As any survey can be compromised by incomprehensible or ambiguous questions, we paid attention to the proper questionnaire design. The following actions were taken to ensure the questionnaire is well-formed and understandable:

- It was divided into 3 separate parts: (1) context information about survey participant's background, (2) assessments of RE/BA techniques' influence on quality characteristics, (3) feedback on questionnaire's understandability and completeness (including open questions about additional RE/BA techniques considered by the participant as crucial for quality characteristics).
- Terms used in the questionnaire were explained and used in a consistent manner. Definitions of quality characteristics and short explanations of RE/BA techniques were associated with each question in which they appeared, to be easily accessible to survey participants.
- A pilot survey involving persons representative for the target group was conducted to verify questionnaire design before the full scale survey started.

The main task of survey participants (included in part 2 of the questionnaire) was to evaluate the impact of particular BA techniques on particular quality

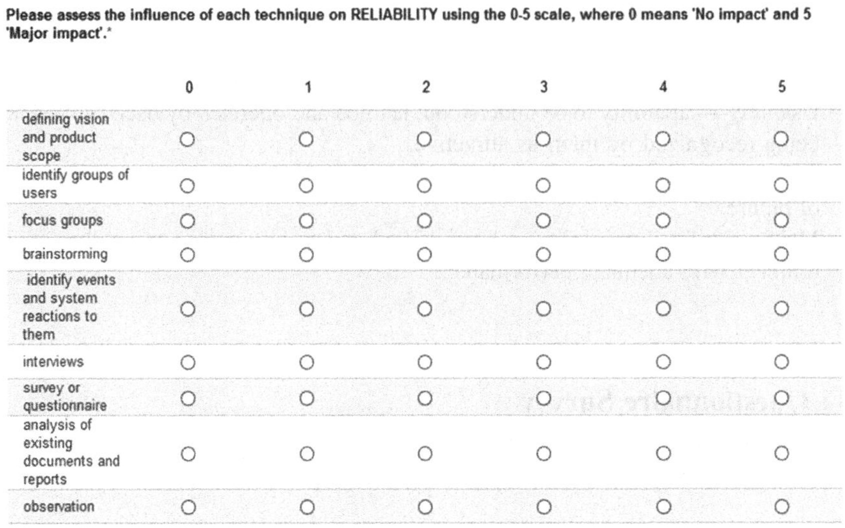

**Please assess the influence of each technique on RELIABILITY using the 0-5 scale, where 0 means 'No impact' and 5 'Major impact'.***

| | 0 | 1 | 2 | 3 | 4 | 5 |
|---|---|---|---|---|---|---|
| defining vision and product scope | O | O | O | O | O | O |
| identify groups of users | O | O | O | O | O | O |
| focus groups | O | O | O | O | O | O |
| brainstorming | O | O | O | O | O | O |
| identify events and system reactions to them | O | O | O | O | O | O |
| interviews | O | O | O | O | O | O |
| survey or questionnaire | O | O | O | O | O | O |
| analysis of existing documents and reports | O | O | O | O | O | O |
| observation | O | O | O | O | O | O |

**Fig. 1** Example question—influence of elicitation techniques on reliability

characteristics using 0–5 Likert scale (0—No impact, 5—Major impact). A partial screenshot from Google Forms demonstrating example of such question is shown in Fig. 1. Each such question focused on techniques from a single area (Elicitation, Validation etc.) and one quality characteristic. Survey participants were supposed to answer 20 questions similar to the one presented in Fig. 1, including in total 124 evaluations (technique/characteristic).

## 4.2 Survey Data Gathering and Processing

The questionnaire was prepared in two language versions (Polish and English) and published online. The survey participants were involved using personal contacts, mailing and publishing invitations on websites dedicated to business analysts and others interested in BA topics. Only the responses with answers addressing all mandatory questions from part 2 were considered. In total 20 such responses were gathered (17 to Polish and 3 to English version). Because of the small number of responses to the English version, we decided not to analyze them separately, but to process all responses together. After a closer look, one response was removed from further consideration (all evaluations were identical starting from some point). The remaining 19 responses were further processed. Context questions from part 1 of the questionnaire provided background information about survey participants, including the following:

- All of them (19) were employed in the IT industry;
- History of professional experience: 1–5 years (9), 6–10 years (6), >10 years (4);
- Current company's number of the employees: 30–120 (10), over 120 (9);
- Current job position: manager/lead (7), designer/developer (5), business analyst (4), architect (1), tester (1), system engineer (1);
- The age demographics: 27–34 years old (11), 18–26 (5), 35–40 (3);

The most important were the answers to part 2 questions—evaluations of RE/BA techniques impact on quality characteristics as perceived by surveyed professionals. We processed these answers, summarized them and visualized as graphs. An example regarding evaluation of Elicitation techniques on Reliability characteristic is shown in Fig. 2 as a diverging stacked bar chart, centered to show how answers about none/weak influence (0–2) and strong (3–5) are distributed. As it is not possible to include full results in the paper, it only shows examples, but the raw data collected in the survey (translated to English) is available at [21].

Table 3 contains the same answers as Fig. 2, but the scale is divided into 2 parts: "positive" 3–5 answers indicating strong influence and "negative 0–2 answers interpreted as weak influence (or none at all).

Distribution of answers for a single assessment was checked to locate its local maximum. As result, a technique could be considered to: have a substantial impact when a local maximum could be found within 3–5 range of values; have a negligible impact if such maximum was in 0–2 range; or remain unclassified if data was inconclusive and neither of two previous conditions were met. There were two such cases for the example presented in Table 3: brainstorming and observation. Such cases were a subject of further analysis and discussion during the interviews conducted later and described in Sect. 4.3.

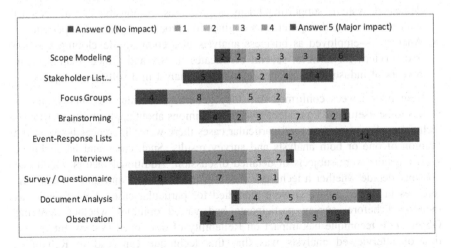

**Fig. 2** Example distribution of answers—impact of elicitation techniques on reliability

**Table 3** Impact of Elicitation techniques on Reliability—answers divided into 2 main categories

| Technique | Influence evaluations (% of answers) | |
|---|---|---|
| | 0–2 values range ("negative") | 3–5 values range ("positive") |
| Scope modeling | 36.9 | 63.1 |
| Stakeholder list, map or personas | 57.9 | 42.1 |
| Focus groups | 89.5 | 10.5 |
| Brainstorming | 47.4 | 52.6 |
| Event-response lists | 0 | 100 |
| Interviews | 84.2 | 15.8 |
| Survey/questionnaire | 94.7 | 5.3 |
| Document analysis | 26.3 | 73.7 |
| Observation | 47.4 | 52.6 |

## 4.3 Interviews

Interviews with two experienced business analysts were arranged to discuss and interpret survey results. We intended to receive general feedback about survey validity and perceived value of its results. We also wished to discuss more thoroughly the cases of inconclusive survey results, hoping to identify the causes of such outcome and additional factors determining the impact we tried to investigate. The interviews with each of 2 analysts were conducted separately. Before the meeting, each one received and reviewed introductory materials. The materials included survey results and the outcome of the analysis we conducted. Our interviewees were:

- Analyst 1—employed as technical lead by a software house which specializes in dedicated systems supporting business processes, responsible for the contact with customers and requirements engineering process, 9 years of experience.
- Analyst 2—employed as business analysts in a company developing systems and services for airlines, several certificates in BA and project management, 6 years of industrial experience as business analyst in 3 software companies.

Both interviewees confirmed that they generally consider the survey and its results to be useful and consistent with their opinions about applicability of RE/BA techniques, however in several particular cases there were differences between the opinion of one or both analysts and survey results. Such cases and inconclusive survey results were subjects of detailed discussion. Additional factors which can possibly decide whether a technique has impact on quality (and thus explain differences in survey answers) were identified for particular cases. For example, as mentioned before, survey participants had varied opinions whether applying Observation technique has impact on Reliability of developed system. Interpretation of interviewed analysts was that this technique can lead to Reliability improvement by identifying user-caused faults, but only when a detailed prototype or pilot deployment is planned in software project. Such comments were noted as additional guidance for RE/BA technique selection.

## 4.4 Final Results

The main result of our research is included in Table 4, which lists the most influential techniques with respect to each of quality characteristics. The table summarizes the results of the survey, confirmed by interviews. The techniques are grouped by RE/BA areas (defined in Sect. 3.1). The ordering within each area (single table row) is meaningful—the most influential technique is listed first.

**Table 4** Most influential RE/BA techniques according to survey results

| Characteristic | Area | Most influential techniques |
| --- | --- | --- |
| Functionality | E | Scope modeling; event-response lists; observation |
| | A | Business model canvas; organizational modeling; prototyping |
| | S | SRS templates |
| | V | Test cases and scenarios; acceptance and evaluation criteria |
| | M | Functional decomposition; lessons learned |
| Usability | E | Observation; focus groups |
| | A | Prototyping; organizational modeling |
| | S | Non-functional requirements analysis |
| | V | Test cases and scenarios; acceptance and evaluation criteria |
| | M | Lessons learned |
| Reliability | E | Event-response lists |
| | A | Risk analysis and mngmt; business model canvas; prototyping |
| | S | Non-functional requirements analysis |
| | V | Test cases and scenarios; reviews |
| | M | Lessons learned; metrics and kpis; functional decomposition |
| Maintainability | E | Scope modeling |
| | A | Business model canvas; data dictionary |
| | S | Non-functional requirements analysis; item tracking |
| | V | Retrospectives; reviews |
| | M | Lessons learned; functional decomposition |

## 5  Validity Threats

This section discusses the different types of threats potentially affecting internal and external validity of our research.

Internal validity concerns the design of research study and potential additional factors that could affect the outcome. In our case, research design could be flawed by wrong selection of RE/BA techniques and/or quality characteristics. Although more techniques are available, described in literature, the systematic approach applied and reliance on renowned sources constitute the argument for including the most important ones. Quality characteristics were derived on the basis of a number of quality models, however the choice of models can be questioned, especially considering that we relied on older sources. More recent models, which reflect the scope of current knowledge on software quality are available e.g. ISO/IEC 25000 series. However, our purpose was to identify a small number of most common characteristics included in established quality models, not to use a complete quality model covering all aspects of e.g. system quality or data quality. Another potential threat is that survey participants lacked sufficient information to answer the questions. We made effort to prevent this by providing them with definitions of all terms used (techniques, characteristics), but nevertheless we asked them (part 3 of the questionnaire) about feedback. The average answer about values on understandability of questions was 3.27 (1–4 Likert scale). Typical threats like history, selection, rivalry or mortality are not applicable due to the method of research (no groups compared, single task of answering questionnaire). Maturation (in particular fatigue) could be an issue—to address this we reviewed data and excluded answers of one participant because of fatigue symptoms.

As for external validity, we made an effort to involve people with an appropriate background (industry practitioners not e.g. undergraduate students as survey respondents, experienced BA as interviewees). However it is difficult to generalize the results because of small size of our sample (19 respondents, 2 interviewees) and the fact that vast majority of them were from a single country.

## 6  Conclusions

In this paper we explored one aspect of RE/BA influence on project outcome by investigating a potential impact of RE/BA techniques on a set of software quality characteristics. For this purpose we conducted a literature search for software quality models, analyzed them and selected a subset of commonly recognized quality characteristics. Also, we identified a representative set of RE/BA techniques recommended by reliable sources. These two sets provided the basis for questionnaire survey and interviews. We conducted a survey, gathered valid responses from 19 industry professionals and conducted 2 interviews with experienced business analysts to discuss and interpret survey results. The main outcome addressing our research question are recommendations regarding techniques to be used in a software project in case a given quality characteristic is considered essential, summarized in Table 4.

A further research is possible by focusing on other RE/BA techniques and/or quality characteristics, by considering more detailed sub-characteristics (e.g. learnability or attractiveness instead of usability) and by identifying additional factors determining whether a given technique is applicable in a given context.

**Acknowledgements** We wish to thank analizait.pl and zarzadzanieit.com websites for publishing information about our survey and all contributors (survey participants and interviewed analysts) for their time and effort. We also thank the anonymous reviewers for improvement suggestions.

# References

1. Boehm, B.W., Brown, J.R., Lipow, M.: Quantitative evaluation of software quality. In: Proceedings of the 2nd International Conference on Software Engineering, pp. 592–605 (1976)
2. Charette, R.N.: Why software fails. IEEE Spectr. **42**(9), 42–49 (2005)
3. Davey, B., Parker, K.: Requirements elicitation problems: a literature analysis. Issues Inform. Sci. Inf. Technol. **12**, 71–82 (2015)
4. dos Santos Soares, M., Cioquetta, D.: Analysis of techniques for documenting user requirements. Comput. Sci. Appl. ICCSA **2012**, 16–28 (2012)
5. Dromey, R.G.: A model for software product quality. IEEE Trans. Softw. Eng. **21**, 146–163 (1995)
6. Ellis, K., Berry, D.: Quantifying the impact of requirements definition and management process maturity on project outcome in large business application development. Requirements Eng. **18**(3), 223–249 (2013)
7. Frączkowski, K., Dabiński, A., Grzesiek, M.: Raport z Polskiego Badania Projektów IT 2010. http://pmresearch.pl/wp-content/downloads/raport_pmresearchpl.pdf (2011)
8. Gorschek, T., Davis, A.: Requirements engineering: in search of the dependent variables. Inf. Softw. Technol. **50**(1), 67–75 (2007)
9. Grady, R.B.: Practical Software Metrics for Project Management and Process Improvement. Prentice Hall, Upper Saddle River (1992)
10. Hofmann, H., Lehner, F.: Requirements engineering as a success factor in software projects. IEEE Softw. **18**(4), 58–66 (2001)
11. Holm, H., Sommestad, T., Bengtsson, J.: Requirements engineering: the quest for the dependent variable. In: 23rd International Requirements Engineering Conference, pp. 16–25 (2015)
12. International Institute of Business Analysis: A Guide to the Business Analysis Body of Knowledge (BABOK Guide) v3 (2015)
13. ISO/IEC: ISO 9126:2001, Software Engineering—Product Quality, Part 1: Quality Model, Geneva (2001)
14. ISO/IEC: ISO 25010:2011, Software Engineering: Software Product Quality Requirements and Evaluation (SQuaRE) Quality Model and Guide, Geneva (2011)
15. Jiang, L., Eberlein, A., Far, B., Mousavi, M.: A methodology for the selection of requirements engineering techniques. Softw. Syst. Model. **7**(3), 303–328 (2008)
16. Khan, H., Asghar, I., Ghayyur, S., Raza, M.: An empirical study of software requirements verification and validation techniques along their mitigation strategies. Asian J. Comput. Inf. Syst. **3**(03) (2015)

17. Kheirkhah, E., Deraman, A.: Important factors in selecting requirements engineering techniques. In: Proceedings of International Symposium on Information Technology, pp. 1–5 (2008)
18. Marciniak, P., Jarzębowicz, A.: An industrial survey on business analysis problems and solutions. In: Proceedings of XVIII KKIO Software Engineering Conference: Software Engineering: Challenges and Solutions, pp. 163–176 (2016)
19. McCall, J.A., Richards, P.K., Walters, G.F.: Factors in software quality: final report. In: Information Systems Programs, General Electric Company (1977)
20. Miquel, J.P., Mauricio, D., Rodríguez, R.: A review of software quality models for the evaluation of software products. Int. J. Softw. Eng. Appl. (IJSEA), 5(6) (2014)
21. Mossakowska, K., Jarzębowicz, A.: Survey Dataset (answers collected). https://drive.google.com/drive/folders/0BwxrBF_-5e_eSlJmSkYxYURDNEk
22. Project Management Institute: A Guide to the Project Management Body of Knowledge (PMBoK), 5th edn. (2013)
23. Project Management Institute: Business Analysis for Practitioners. A Practice Guide (2015)
24. Radliński, Ł.: Empirical analysis of the impact of requirements engineering on software quality. In: International Working Conference on Requirements Engineering: Foundation for Software Quality, pp. 232–238 (2012)
25. Radliński, Ł.: How software development factors influence user satisfaction in meeting business objectives and requirements? XVI KKIO Software Engineering Conference: Software Engineering from Research and Practice Perspectives, Nakom, pp. 101–119 (2014)
26. Sethia, N.K., Pillai, A.S.: A study on the software requirements elicitation issues—its causes and effects. In: Proceedings of World Congress on Information and Communication Technologies, pp. 245–252 (2013)
27. Sommerville, I., Ransom, J.: An empirical study of industrial requirements engineering process assessment and improvement. ACM Trans. Softw. Eng. Methodol. 14(1), 85–117 (2005)
28. Thapar, S.S., Singh, P., Rani, S.: Challenges to development of standard software quality model. Int. J. Comput. Appl. 49(10) (2012)
29. The Standish Group International: Chaos Report 2014 (2014)
30. Wellsandt, S., Hribernik, K., Thoben, K.: Qualitative comparison of requirements elicitation techniques that are used to collect feedback information about product use. In: Proceedings of 24th CIRP Design Conference, pp. 212–217 (2014)
31. Wiegers, K., Beatty, J.: Software Requirements, 3rd edn. Microsoft Press (2013)

# Female Leadership in Software Projects— A Preliminary Result on Leadership Style and Project Context Factors

Anh Nguyen-Duc, Soudabeh Khodambashi, Jon Atle Gulla,
John Krogstie and Pekka Abrahamsson

**Abstract** Women have been shown to be effective leaders in many team-based situations. However, it is also well-recognized that women are underrepresented in engineering and technology areas, which leads to wasted efforts and a lack of diversity in professional organizations. Although studies about gender and leadership are rich, research focusing on engineering-specific activities, are scarce. To react on this gap, we explored the experience of female leaders of software development projects and possible context factors that influence leadership effectiveness. The study was conducted as a longitudinal multiple case study. Data was collected from survey, interviews, observation and project reports. In this work, we reported some preliminary findings related to leadership style, team perception on leadership and team-task context factors. We found a strong correlation between perceived team leadership and task management. We also observed a potential association between human-oriented leading approach in low customer involvement scenarios and task-oriented leading approach in high customer involvement situations.

A. Nguyen-Duc (✉) · S. Khodambashi · J.A. Gulla · J. Krogstie · P. Abrahamsson
Department of Computer and Information Science, Norwegian University of Science
and Technology, Trondheim, Norway
e-mail: anhn@ntnu.no

S. Khodambashi
e-mail: soudabeh@ntnu.no

J.A. Gulla
e-mail: jon.atle.gulla@ntnu.no

J. Krogstie
e-mail: john.krogstie@ntnu.no

P. Abrahamsson
e-mail: pekkaa@ntnu.no

A. Nguyen-Duc
Department of Business Administration and Computer Science,
University College of Southeast Norway, Notodden, Norway

P. Abrahamsson
Faculty of Information Technology, University of Jyväskylä, Jyväskylä, Finland

© Springer International Publishing AG 2018
P. Kosiuczenko and L. Madeyski (eds.), *Towards a Synergistic Combination
of Research and Practice in Software Engineering*, Studies in Computational
Intelligence 733, DOI 10.1007/978-3-319-65208-5_11

149

**Keywords** Female leadership · Software engineering · Contingency model · Project contextual factors · Team coordination · Team performance

# 1 Introduction

The rise of female workforce has recently gained attention as an upcoming paradigm shift in modern organizations and professional teams. Women have been found to be effective leaders in team-based, consensually-driven organizational structures that are becoming more and more popular [1, 2]. In STEM (Science, Technology, Engineering and Mathematics), however, women are still under-representative. Averaged across the globe, women accounted for less than a third (28.4%) of those employed in scientific research and development (R&D) in 2013 [3]. Although women hold close to 45% all jobs in the U.S. economy, they take less than 25% of STEM jobs [4]. The underrepresentation of women in STEM is problematic as it results in the waste of talents, gender diversity and creativity in professional workplaces.

Despite of a rich literature on gender studies, research seems still inconclusive about the reasons for the underrepresentation of women in STEM, and more importantly, how to deal with that. Many scholars believe that the existence of gender stereotypes is one of the main reasons [5–7], as the traditional male-biased practices and leadership norms function to exclude women [8]. There is an increasing attention to the gender topic in IT, i.e. Software Engineering (SE) area. A search in the Scopus database at 21st March, 2017 gave us 88 peer-reviewed publications about gender-related topics in SE area. This number is very modest compared to gender studies in other fields, i.e. Business and Management.

With the important role of software in modern informational and cyber-physical products, understanding human factors of software development, including the gender perspective, is essential to design a processes, practices and guidelines that best utilize the available workforces. To react on this above gap, we conducted a research on female leadership in software development projects, with the aim of understanding the relationship between SE-specific context factors, gender stereotype and leadership. Inheriting from a body of knowledge on gender and leadership, we do not focus on differentiating male and female leadership behaviors. Instead, we would like to know if there are project situations that best supports female leadership. Our general research question is: **How can we characterize the female leadership under different software project situations?** The paper presents our research design and preliminary results for analyzing women leadership styles and their leading situations in term of team and task characteristics.

The paper is organized as follows: Sect. 2 briefly presents background on leadership and gender studies, and state-of-the-art on gender study on SE. While Sect. 3 presents research methodology, Sect. 4 describes our preliminary findings and discussion. Section 5 is the conclusion and future work.

# 2   Background

## 2.1   Gender and Leadership

**Leadership**: a clear definition of what "leadership" remains somewhat elusive as numerous definitions of "leadership" has been found in literature. We found some definitions that apply well to understand female leadership in our context setting: *"behavior of an individual ... directing the activities of a group toward a shared goal"* [9], and *"... ability of an individual to influence, motivate, and enable others to contribute toward the effectiveness and success of the organization ..."* [10].

**Gender stereotyping**: research has produced various theories about whether leadership traits and behaviors differ between men and women as distinctive bio-logical groups. First of all, one needs to understand gender a multidimensional personality characteristic most commonly described by the difference in term of (1) biology and sex, (2) gender role, (3) causal factors and (4) attitudinal drivers [11]. Gender stereotypes are somewhat culturally shared beliefs that dictate expectations about how women and men are and how they ought to behave [12]. Thus, stereotypes can be both descriptive and prescriptive in nature. Perceived feminine competences, which are required in STEM, include, for instance, com-munication competences, customer and workplace relationship competences, and creativity [13, 14]. Technical competences, such as programming software archi-tecture, are perceived as being fundamentally attached to male [15, 16].

**Leadership style**: such as relationship-oriented and task-oriented styles [17], transformational and transactional styles [1, 11, 18], and directive and participative leadership [19] have been an important topic in literature of gender and leader-ship. Contingency approach suggests that not only the traits and behaviors of a leader can explain for leadership effectiveness, but also the situation that the leader is at [17, 20, 21]. Fiedler developed the LPC Contingency Model, which focuses on the relationship between a trait termed the "least preferred coworker" (LPC) score and leadership effectiveness. He concluded that the most favorable situations for leaders were those in which they were well liked (good leader-member relations), directed a well-defined job (high task structure), and had a powerful position (high position power) [17, 20, 21].

**Gender and leadership**: other style category describes gender difference in term of transformational and transactional leadership [22]. Transformational leaders are characterized as inspiring, motivating, being attentive to and intellectually chal-lenging their followers, where transactional leaders are described as contractual, corrective, and critical in their interactions with employees [22]. Female leaders were found to be more transformational than male leaders and also engaged in more of the contingent reward behaviors that are a component of transactional leadership [1]. Overall, researchers have asserted that there is no *"one style fits all"* solution to leadership issues and that the efficacy of various styles is contextual [23, 24]. While there are some differences between men and women when it comes to style, these differences do not lead to a clear advantage of either gender across contexts [18].

## 2.2  Gender Lens in SE

SE research on gender is heavily driven by theories of gender stereotyping. In requirement engineering tasks, female performances are associated with personal factors. Less success is expected and achievements are less attributed to the own abilities [25]. Multiple case studies on women in course-based software projects characterize the collaborative learning environments for women participating [26]. The authors identified four common themes: working with others; productivity; confidence; and interest in IT careers. One key finding was that collaboration, emerging from face-to-face meetings, helps female students to build confidence via higher quality products and to reduce amount of time spent on assignments [26].

SE research also focuses on how men and women work differently regarding to specialized SE tasks. In term of coding, it is found to be different compatibility and communication levels between same gender pair and mixed gender pair [27]. Women were found to often develop deficient elements and inappropriate strategies in complex problem solving. Women tend to use bottom-up strategies while men are more risk prone and use more top-down strategies [28]. Regarding to quality assurance and testing, female developers expressed a lower level of self-efficacy than males did about their abilities to debug [29]. Further, women were less likely than males were to accept the new debugging features. Although there was no gender difference in fixing the seeded bugs, women introduced more new bugs—which remained unfixed. There was also evidence on the difference between programming environments as to which features men and women use and explore [30].

## 3  Research Approach

### 3.1  Conceptual Framework

Starting from the demand of understanding women's participation in SE, we gathered several factors that were found to be relevant to leadership effectiveness and team performance, as illustrated in Fig. 1. The factors were found from literature as influencing factors to team performance. Based on the situational leadership approaches [17, 20], we identify factors related to leaders' characteristics:

- **Leadership style**: leadership styles, such as human-oriented style and task-oriented style [17], transformational and transactional styles [1, 11, 18], and directive and participative leadership [19] is an important variable in literature of gender and leadership. In this study, we do not focus on exploring the difference of leadership styles between women and men as it has been extensively researched. To simplify the identification of leadership style, we adopted the Fiedler LPC Contingency model [17, 20, 21].
- **Leaders' experience**: personal characteristics, such as females' interest in IT careers [26], previous experience with leadership [31], personality [31] can

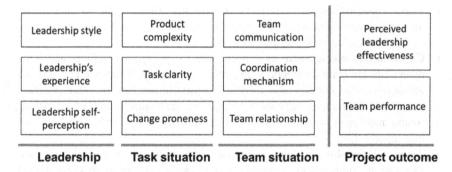

| Leadership style | Product complexity | Team communication | Perceived leadership effectiveness |
| Leadership's experience | Task clarity | Coordination mechanism | |
| Leadership self-perception | Change proneness | Team relationship | Team performance |
| **Leadership** | **Task situation** | **Team situation** | **Project outcome** |

**Fig. 1** The conceptual framework

influence the leadership effectiveness in a given situation. It is found that if women have no task-specific experience, they do not feel confident while males apply general knowledge for specific tasks [28]. We expect to gather as much information about leadership's characteristics as possible via interviews and informal discussions.

- **Leadership self-perception** on her ability of decision making, inter-personal management and task management could give a proxy to understand leadership behavior and indirectly team performance.

Research has also shown that women are more likely to be in problematic organizational circumstances [32], as in smooth situations, agentic characteristics mattered more for leader selection, whereas in times of crisis, interpersonal attributes were deemed more important [32]. To relate the problematic situation in SE project context, we identify the product complexity, task clarity and change proneness as they have been used in literature as proxies for project context factors:

- **Product complexity**: IEEE standard defines software complexity as *"the degree to which a system or component has a design or implementation that is difficult to understand and verify"* [33]. The complexity reflects how difficult to understand the requirement, and to provide the solution.
- **Task clarity**: ambiguous requirements can lead to different interpretations among developers that lead to mistakes and confusion and consequently wasted efforts. Requirement clarity describes how clearly are requirements are presented to and understood among project stakeholders in the early phases of projects.
- **Change proneness**: characterizes how likely a requirement will change over time. For a non-experience development team, frequent changes might introduce difficulty in completing requirements and achieving customer satisfaction.

Learning from a previous work [26], we identify team dynamic as an important factor to understand female leadership behavior.

- **Team communication**: can occur in various forms, such as face-to-face meetings, pair programming and social conversations. The frequency, richness of communication is within our concern.
- **Coordination mechanism**: as a mechanism to synchronize activity and information among team members, is also related to team performance [34, 35]. We identify whether coordination mechanism used is mechanistic (task organization, task assignment, schedules, plans, project controls and specifications, routine meetings) or organic (informal discussions, developer's comments, and bug reports) [36]
- **Team relationship**: as female leadership is commonly perceived with communication competences, customer and workplace relationship competences, and creativity [13, 14], it is important to understand team members, external stakeholders and their relationships to team leaders.

**Team performance and perceived leadership effectiveness**: we identify both subjective team performance (feedbacks from team members and leader herself about how good are they as a team) and objective team performance (external evaluation of teamwork as project outcomes). Team performance evaluation is often used as a proxy of leader's effectiveness [18]. Besides, female leadership might be perceived differently due to prejudice and discrimination directed against women as leaders [2].

## 3.2 Study Design

To explore the prototyping practices, we conducted a multiple exploratory case study [37]. According to Yin [37], a case study design is appropriate when (a) the focus of the study is to answer "how" and "why" questions; (b) there is probably high influence of contextual factors on the studied phenomenon. Exploratory case studies are suitable to explain the presumed causal links in real-life interventions. A multiple case study enables the researcher to explore differences within and between cases.

The underlying cases were based on software development projects in the Customer Driven Project course at the Norwegian University of Science and Technology (NTNU). Students in the course were randomly divided into groups of five to seven members. Each group has to carry out a three-month-long project for real customers from software companies, governmental agencies or research institutes. The project idea reflects a recognized need from the customer and can lead to the development of a new solution or a component of an existing one. The goal is to give the students practical experience in carrying out project development and management activities in a context as close to industry as possible. A typical project involves customers, end-users, advisors from the university and any relevant third parties. The success of the project is evaluated by a board of a course responsible, an external examiner and a team supervisor. The evaluation criteria are based on the quality of the deliverables, customer satisfaction and teamwork.

To establish the intervention, we put a female student as a leader of each project team. In Fall 2015, we observed 13 female-lead teams of 79 students. Since each student spends about 20 h per week on the project in one semester, a typical project of 6 students corresponds to about 2,000 person hours. In the scope of this study, we will present research conducted in Fall 2015.

## 3.3  Data Collection

The data collection instrument was designed to cover the conceptual elements, as described in Table 1. Data source triangulation was ensured by using both qualitative (interviews, documents, observation) and quantitative data (survey).

**Instrument 1: Leadership style survey**. We utilized the Fiedler LPC questionnaire to identify the leadership style [38]. The survey design is based on Fielder contingency model, which proposed that a leader's effectiveness is based on the match between "leadership style" and "situational favorableness" [17]. The survey responses were collected in the first two weeks of the project.

**Instrument 2: Project plan**. Within the course setting, each team provided their project plan with descriptions of team structure, roles, and preliminary study about the product requirements and time plan.

**Instrument 3: Team meeting observation notes**. Each team was assigned a supervisor who will assist the team from a coursework perspective. The supervisor meeting with teams offered an opportunity for research team to observe team behavior during meetings. Observation focused on team's ability to perform collective problem solving, relationship among team members and with team leader and team collaboration and coordination practices.

**Instrument 4: Leadership performance survey**. We designed a second survey to collect team's perception on their own teamwork and leadership. The survey used a five-point Likert scale to collect leaders and team member's opinions on (1) their own performance, (2) collective decision making, (3) team leadership and (4) task management practices.

**Table 1** Matrix of conceptual elements and collection instrument

| Conceptual element/Instrument | I1 | I2 | I3 | I4 | I5 | I6 |
|---|---|---|---|---|---|---|
| Leadership style | ✓ |  | ✓ |  | ✓ |  |
| Leaders self reflection |  |  |  | ✓ | ✓ |  |
| Leader's experience |  |  |  |  | ✓ |  |
| Task clarity |  | ✓ | ✓ |  |  | ✓ |
| Change proneness |  |  | ✓ |  |  | ✓ |
| Team collaboration |  |  | ✓ | ✓ | ✓ | ✓ |
| Team relationship |  | ✓ | ✓ | ✓ | ✓ | ✓ |
| Team communication |  |  | ✓ | ✓ | ✓ | ✓ |
| Team coordination |  |  | ✓ |  | ✓ | ✓ |
| Perceived team performance |  |  |  | ✓ |  |  |

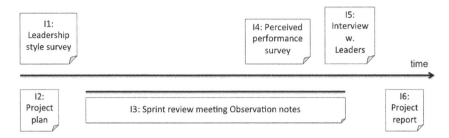

**Fig. 2** Data collection process

**Instrument 5: Leader interview**. Interview with team leader was a valuable data to understand in-depth about the perception of team leader about the team performance, their experience on leading and working in SE tasks. The interview was designed as semi-structured interview that allows surprise development of the interview scenario to interesting results.

**Instrument 6**: **Final project report**. Each team delivered a 150–200 pages project report describing project planning and management, product requirement and architecture, testing and delivery. Especially, a final part of the report consisted of team reflection on project mandate, teamwork and supervision.

The process of data collection was described as in Fig. 2. The data collection periods was done from Aug 2015 to Nov 2015. While I1 and I4 were mainly done by the first authors, I3 and I5 were jointly collected by all co-authors and other supervisors in the course. I2 and I6 were collected as a part of the course delivery.

## 3.4 *Data Analysis*

At this phase, we limited ourselves in simply qualitative and quantitative analysis. The major part of analysis was done by the first author and revised by the rest.

**Narrative analysis**: is a simple form of qualitative analyzing data from i.e. interview transcripts, observation notes and textual descriptions [20]. Narratives or stories occur when one or more speakers engage in sharing an experience, which is suitable material for abstracting the project experience from our team leaders. We performed a simplified version of the analysis to extract from each project story categorical information. For instance, the level of team structure and task structure in the project will be extracted and interpreted from project report (I6). In the end of this step, we came up with a list of ordinal variables representing project context factors.

**Descriptive analysis**: quantitatively summarize characteristics of project, team and leader of our sample. The statistical summary was provided for leadership style (Sect. 4.1) and team reflection on leadership (Sect. 4.3).

**Correlation analysis**: investigates the extent to which changes in the value of an attribute are associated with changes in another attribute. Though a correlation between two variables does not necessarily result in causal effect [39], it is still an effective method to select candidate variables for causation relationship. We find Spearman's rank correlation is suitable in our case as the test does not have any assumptions about the distribution of the data and is the appropriate correlation analysis in case of ordinal variables.

## 4 Preliminary Results and Discussion

### 4.1 Leadership Style

In our sample, the LPC score ranges from 47 to 109, with median value is 67, as shown in Fig. 3. An interpretation approach considers the score 57 or below as a task-oriented leader and score 64 to above as a relationship-oriented leader [48]. Consequently, in general, our sample biases towards relationship-oriented leadership style. We identified five groups with task-oriented leaders, six groups with relationship-oriented leaders and two groups with mixed leaders.

We preliminarily validate the relationship between female leadership style alone and the objective project performance:

*H0: The LPC leadership style and objective project performance are independent.*

*Test result: X-squared = 6.3254, df = 12, p-value = 0.8988.*

**Fig. 3** Boxplot of female leaders' LPC score

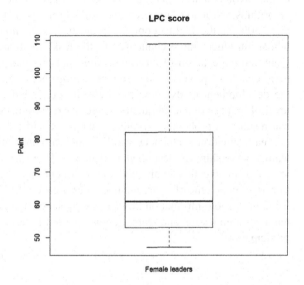

**Table 2** Task-related contextual factors

| Product complexity | Task ambiguity | Change proneness | No. of project |
|---|---|---|---|
| High | Medium | Low | 3 |
| High | Low | High | 1 |
| High | Low | Medium | 1 |
| Medium | High | Medium | 1 |
| Medium | Low | Medium | 1 |
| Medium | Medium | Low | 1 |
| Low | High | Medium | 1 |
| Low | Medium | Medium | 1 |
| Low | Low | High | 1 |
| Low | Low | Low | 1 |

The result of a Chi square test of independence between the two variables did not allow us to reject the null hypothesis, which infers no direct relationship between leadership style and objective team performance. This somehow aligns with our expectation, as the leadership effectiveness should be impacted by a complex combination among the leadership style and the situational favorableness Table 2.

## 4.2  Project Contextual Situation

Project contextual situation is characterized by task situation and team situation. Task situation is characterized by product complexity, task clarity and change proneness, as described in Fig. 1. Six projects were found to be highly complex due to the involvement of (1) multiple set of hardware devices, (2) machine learning algorithms, (3) market research and validation, and (4) industry specific technologies. 50% of the highly complex projects has medium level of task ambiguity, which introduces more challenges to the task situation. In such case, the requirement and expectation of customers were not clarified to all team members in the early phases of projects. The remaining projects (seven projects) typically involve the development of dynamic mobile/web applications. It is noticed that the high level of ambiguity and frequent changes of requirement in later phases of projects can result in a problematic project situation regardless of product complexity.

Team situation is characterized by team communication, team coordination and team relationship, as described in Table 3. There are three projects found in as smooth team situations with good amount of communication among team members. Furthermore, (some of) team members knew each other before or quickly were able to set up a good information flow. There is one challenging team situation with low team communication and team members varied in term of project commitment and participation.

**Table 3** Team-related contextual factors

| Team communication | Coordination mechanism | Team relationship | No. of projects |
| --- | --- | --- | --- |
| High | Mechanistic | High | 1 |
| High | Mechanistic | Medium | 2 |
| High | Organic | High | 2 |
| High | Organic | Medium | 2 |
| Medium | Mechanistic | Medium | 1 |
| Medium | Organic | Medium | 3 |
| Medium | Organic | High | 1 |
| Low | Mechanistic | Medium | 1 |

## 4.3 Correlation Analysis

We calculated the Spearman correlation coefficient value among variables, as shown in Fig. 4. In the figure, the sizes of circles represent the coefficient value, and the color (blue or red) represents the influence direction (positive or negative). Interestingly, we found no correlation between objective team performance and perceived team performance. Leadership style is not related to team's perception on their performances or team's perception on their leadership effectiveness. Perceived team leadership is also not correlated with task.

We are interested in female leadership correlations that are significant at 0.1 level, as shown in Table 4. Hopkins calls a correlation coefficient value between 0.5–0.7 large, 0.7–0.9 very large, and 0.9–1.0 almost perfect [40]. According to this scheme, objective team performance is largely correlated with team communication. The team with higher grade seems to have a better communication, in term of quality, frequency and effort spending on working together.

**Fig. 4** Spearman correlation analysis of project context factors

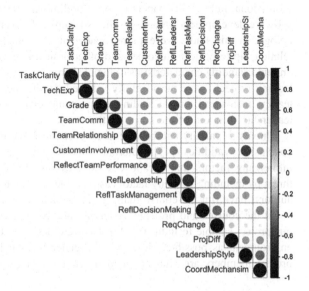

**Table 4** Significant correlational relationships with female leadership

| Variable 1 | Variable 2 | Spearman Rho | P value |
| --- | --- | --- | --- |
| ObjTeamPerformance | TeamComm | 0.61 | 0.02 |
| ReflLeadership | ObjTeamPerformance | 0.50 | 0.08 |
| ReflLeadership | ReflTaskManagement | 0.53 | 0.05 |
| LeadershipStyle | CustomerInvolvement | −0.62 | 0.02 |
| LeadershipStyle | CoordMechansim | 0.50 | 0.07 |

As shown in Table 4, team perception on team leadership is correlated with objective performance, but also with their perception on how tasks are managed. Teams with higher grades tend to be teams having positive experience with team leadership.

In the context of female leading projects, we found a strong correlation between the leadership style and customer involvement and coordination mechanisms. This of course does not exclude the cases of male leading projects. The negative coefficient value means that, in one hand, the lower score of leadership style, representing task-oriented approach is probably associated with situations of high level of customer involvement. In the other hand, the higher score of leadership style, which is human-oriented approach, can be associated with situations of low customer involvement. However, the correlation can occur because of the current customer involvement situation and assigned leadership.

## 5 Conclusions and Future Work

Female participation in software industry is far less representative than that in other professional fields. This papers report a preliminary result from 13 female leading software development projects in the Norwegian University of Science and Technology. In this study, we explored leadership's characteristics and their potential relationship to project outcomes and other project context factors. Given the focus on understanding characteristics of female behavior in project leadership, the female leading projects were purposefully selected. While the comparison between male leading teams and female leading teams could give a comparable observation on leadership, it is out of the scope of this work.

Threats to validity can be discussed in term of internal, external, construct and conclusion validities [41]. Internal validity can occur when there are other influencing factors that escape our observation. We attempted to eliminate this threat by including as many factors as possible. The course coordinators (also paper authors) have run the course many years, which gave us a comprehensive insight about the courses and projects. External validity refers to the ability to generalize our project sample to industry. Projects were designed to be as close to industrial environment as possible. However, the generalization is limited by the fact that project members are students who did not spend full-time on the project. The female leadership position is also an experimented situation that might not happen naturally in industry. Construct validity relates to the transformation of conceptual elements to variables. The conceptual elements were constructed from existing literature. We also borrowed measures that were successfully adopted in literature, i.e. LPC score or team reflection survey. The transformation of textual information from data source, i.e. project reports, was cross-checked by co-authors of the paper, to reduce the bias in construct validity.

Our preliminary observation shows that the perception on team leadership is probably impacted by how the team perceive on task management. Perceived leadership and objective team performance is largely correlated, inferring one can be the indicator of the other. We also confirm that team communication is crucial for project success [34]. Last but not least, we observed a potential association between human-oriented leading approach in low customer involvement scenarios and task-oriented leading approach in high customer involvement situations. Considering the limited size of our sample, the results are not conclusive at this stage. In the future work, we will run a similar study with other customer projects to compare the results with the current study. We also plan to run different regression analysis on the whole dataset to identify indicators of female leadership effectiveness. Moreover, studying interview data with female leaders (I5) would give insight and explanation about relationships between female leadership and project factors. After having insight on female leadership in software projects, the observation will be compared and contrast with those in male leading projects.

**Acknowledgements** We would like to express our appreciation for Katja Abrahamsson, Francesco Valerio Gianni, Simone Mora, and Juhani Risku for participating in data collection.

# References

1. Eagly, A.H., Johannesen-Schmidt, M.C., Van Engen, M.L.: Transformational, transactional, and Laissez-Faire leadership styles: a meta-analysis comparing women and men. Psychol. Bull. **129**, 569–591 (2003)
2. Eagly, A.H., Carli, L.L.: The female leadership advantage: an evaluation of the evidence. Leadership Q **14**, 807–834, 12 (2003)
3. Women In Science, Technology, Engineering, and Mathematics. http://www.catalyst.org/knowledge/women-science-technology-engineering-and-mathematics-stem
4. Women in STEM: A Gender Gap to Innovation. http://www.esa.doc.gov/sites/default/files/womeninstemagaptoinnovation8311.pdf
5. Wajcman, J.: Feminism Confronts Technology. Polity, Cambridge (1991)
6. Robertson, M., Newell, S., Swan, J.: The issue of gender within computing: reflections from the UK and Scandinavia. Inf. Syst. J. **11**(2), 111–126 (2011)
7. Wilson, M.: A conceptual framework for studying gender in information systems research. J. Inf. Technol. **19**(1), 81–92 (2004)
8. Dunn, D., Gerlach, J.M., Hyle, A.E.: Gender and leadership: reactions of women in higher education administration. Int. J. Leadership Change **2** (2014)

9. Hemphill, J.K., Coons, A.E.: Development of the leader behavior description questionnaire. In: Stodgill, R.M., Coons, A.E. (eds.) Leader Behavior: Its Description and Measurement, pp. 6–38. Bureau of Business Research, The Ohio State University, Columbus, OH (1957)
10. Yukl, G.: Leadership in Organizations, 5th edn. Prentice Hall, Upper Saddle River, NJ (2002)
11. Appelbaum, S.H., Audet, L., Miller, J.C.: Gender and leadership? Leadership and gender? A journey through the landscape of theories. J. Leadership Organ. Dev. **24**, 43–51 (2003)
12. Eagly, A.H., Karau, S.J.: Role congruity theory of prejudice toward female leaders. Psychol. Rev. **109**, 573–598 (2000)
13. von Hellens, L., Nielsen, S., Beekhuyzen, J.: An exploration of dualisms in female perceptions of IT work. J. Inf. Technol. Educ. **3**, 103–116 (2004)
14. Trauth, E.M., Cain, C.C., Joshi, K.D.: Embracing intersectionality in gender and IT career choice research. In: 50th Conference on Computers and People Research, pp. 199–212 (2012)
15. Adam, A., Griffiths, M., Keogh, C.: Being an "it" in IT: gendered identities in IT work. Eur. J. Inf. Syst. **15**(4), 368–378 (2006)
16. Henwood, F., Plumeridge, S., Stepulevage, L.: A tale of two cultures? Gender and inequality in computer education. In: Wyatt, S., Henwood, F., Miller, N., Senker, P. (eds.) Technology and In/Equality: Questioning the Information Society, pp. 111–128. Routledge (2000)
17. Fiedler, F.E.: Leader Attitudes and Group Effectiveness, Urbana. University of Illinois Press, IL (1958)
18. Gipson, A.N., Pfaff, D.L., Mendelsohn, D.B., Catenacci, L.T., Burke, W.W.: Women and leadership: selection, development, leadership style, and performance. J. Appl. Behav. Sci. **53**, 32–65 (2017)
19. Koopman, P.L., Wierdsma, A.F.M.: Participative management. In: Doentu, P.J.D., Thierry, H., de-Wolf, C.J. (eds.) Personnel Psychology: Handbook of Work And Organizational Psychology, vol. 3, pp. 297–324 (1998)
20. Riessman, C.K.: Narrative Analysis. Sage Publications, Newbury Park (1993)
21. Fiedler, F.E.: A contingency model of leadership effectiveness. In: Berkowitz, L. (ed.) Advances in Experimental Social Psychology, pp. 149–190. Academic Press, New York, NY (1964)
22. Bass, B.M.: From transactional to transformational leadership: learning to share the vision. Org. Dyn. **18**(3), 19–31 (1990)
23. van Engen, M.L., Willemsen, T.M.: Sex and leadership styles: a meta-analysis of research published in the 1990s. Psychol. Rep. **94**, 3–18 (2004)
24. Eagly, A.H.: Women as leaders: leadership style versus leaders' values and attitudes. In: Gender and Work: Challenging Conventional Wisdom, pp. 4–11. Boston, MA (2013)
25. Gramß, D., Frank, T., Rehberger, S., Vogel-Heuser, B.: Female characteristics and requirements in software engineering in mechanical engineering. In: International Conference on Interactive Collaborative Learning, ICL 2014, pp. 272–279 (2014)
26. Berenson, S.B., Slaten, K.M., Williams, L., Ho, C.W.: Voices of women in a software engineering course: reflections on collaboration. ACM J. Educ. Resour. Comput. **4** (2004)
27. Choi, K.S.: A comparative analysis of different gender pair combinations in pair programming. Behav. Inf. Technol. **34**, 825–837 (2015)
28. Fisher, M., Cox, A., Zhao, L.: Using sex differences to link spatial cognition and program comprehension software maintenance. In: 22nd IEEE International Conference on Software Maintenance, pp. 289–298 (2006)
29. Beckwith, L., Kissinger, C., Burnett, M., Wiedenbeck, S., Lawrance, J., Blackwell, A., et al.: Tinkering and gender in end-user programmers' debugging. In: Conference on Human Factors in Computing Systems, pp. 231–240 (2006)
30. Burnett, M., Fleming, S.D., Iqbal, S., Venolia, G., Rajaram, V., Farooq, U., et al.: Gender differences and programming environments: across programming populations. In: The 2010 ACM-IEEE International Symposium on Empirical Software Engineering and Measurement (2010)
31. Hogan, R., Curphy, G.J., Hogan, J.: What do we know about leadership, effectiveness and personality. Am. Psychol. **49**(6), 493–504 (1994)

32. Ryan, M.K., Haslam, S.A., Hersby, M.D., Bongiorno, R.: Think crisis–think female: the glass cliff and contextual variation in the think manager–think male stereotype. J. Appl. Psychol. **96**, 470–484 (2011)
33. IEEE Standard Glossary of Software Engineering Terminology, report IEEE Std 610.12-1990
34. Nguyen-Duc, A., Cruzes, D.S., Conradi, R.: The impact of global dispersion on coordination, team performance and software quality—a systematic literature review. Inf. Softw. Technol. **57**, 277–294 (2015)
35. Nguyen-Duc, A., Mockus, A., Hackbarth, R., Palframan, J.: Forking and coordination in multi-platform development: a case study. In: The ACM/ IEEE International Symposium on Empirical Software Engineering and Measurement (ESEM), Torino, Italy (2014)
36. Thompson, J.D.: Organizations in Action. McGraw-Hill (1967)
37. Yin, R.K.: Case Study Research: Design and Methods (Applied Social Research Methods), 5th edn. SAGE Publications, Inc. (2014)
38. Cohen, J.: Statistical Power Analysis for the Behavioral Sciences, 2nd edn. Routledge Academic (1988)
39. W. G. Hopkins, A new view of statistics. Sport Science. http://www.sportsci.org/resource/stats (2003)
40. Runeson, P., Höst, M.: Guidelines for conducting and reporting case study research in software engineering. Empirical Softw. Eng. **14**(2), 131–164 (2009)
41. Trauth, E.M.: The role of theory in gender and information systems research. Inf. Organ. **23** (4), 277–293 (2013)

# Modelling and Verification of Real-Time Systems with Alvis

Marcin Szpyrka, Łukasz Podolski and Michał Wypych

**Abstract** Alvis is a formal language specifically intended for modelling systems consisting of concurrently operating units. By default, the time dependencies of the modelled system are taken into account, what is expressed by the possibility of determining the duration of each statement performed by the model components. This makes Alvis suitable for modelling and verification of real-time systems. The paper focuses on Alvis time models. The article outlines the syntax and semantics of such models, and discusses the main issues related to the generation of Labelled Transition Systems for time models. Particular attention was paid to tools that support the verification process.

## 1 Introduction

The development of concurrent systems for which we want to guarantee a high level of reliability can be a tedious and difficult task. High degree of concurrency makes system more flexible but increases the risk of leaving in such a system significant bugs that cannot be detected during the system testing stage. Combination of concurrency and time dependencies makes the task even more difficult. The verification and validation process for such systems cannot be based on classical approaches like peer reviewing and testing [5]. Application of formal methods in the development process may significantly reduce the costs and affects the product quality [2]. However, the use of formal methods in the development process requires additional

M. Szpyrka (✉) · Ł. Podolski · M. Wypych
Department of Applied Computer Science, AGH University of Science and Technology,
Al. Mickiewicza 30, 30-059 Krakow, Poland
e-mail: mszpyrka@agh.edu.pl

Ł. Podolski
e-mail: podolski@agh.edu.pl

M. Wypych
e-mail: mwypych@agh.edu.pl

© Springer International Publishing AG 2018
P. Kosiuczenko and L. Madeyski (eds.), *Towards a Synergistic Combination of Research and Practice in Software Engineering*, Studies in Computational Intelligence 733, DOI 10.1007/978-3-319-65208-5_12

effort in learning new skills and spending more time on analysis and design stages of a software development cycle.

The most popular formal languages that can be used for modelling real-time systems include selected classes of Petri nets [9, 13, 14], time automata [3], and time process algebras [1]. Due to their specific mathematical syntax, these languages are usually treated as the ones suitable only for scientists. In contrast to these languages Alvis [16], [15] is being developed for making the modelling and verification process simpler and more accessible to software developers. The heavy mathematical foundations are hidden from users without compromising the capabilities and expressive power of the formalism. Alvis is equipped with a graphical language [15] for modelling communication channels between the considered system units (called *agents* in Alvis) and a high level programming language for defining agents's behaviour [18]. The language is supported by a set of tools called *Alvis Toolkit*. The software can be used for designing Alvis models, for generating executable Haskell files, for generating Labelled Transition Systems [2] (LTS graphs), and for exporting the LTS graphs to DOT, Aldebaran or CSV formats. This makes possible to verify Alvis models with the most popular model checkers including nuXmv [6] and CADP [7].

The paper is organised as follows. Section 2 provides a short introduction to the Alvis language. Section 3 deals with semantics of time Alvis models. The concept of LTS graphs for time models is discussed in Sect. 4. Conclusions and future works are presented in the final section.

## 2    Alvis Language in a Nutshell

An Alvis model is a system of *agents* that usually run concurrently, communicate with each other, compete for shared resources, etc. The set of agents can be divided into two subsets *active agents* and *passive agents*. Active agents can be treated as processes. Current version of the Alvis models (time and non-time models) supports only so-called $\alpha^0$ system layer. It means that each active agent has access to its own processor and can perform its statements in parallel with other agents. There is also $\alpha^1$ system layer under development. The later layer is based on the assumption that there is only one processor and all active agents compete for access to the processor.

Passive agents are used to represent shared resources. They provide a set of services for other agents and prevent simultaneous access to the data they store. Passive agents' services do not have their own thread of control but always work in the context of an active agent.

The set of agents of a given model is described with two description layers. The graphical layer is called *communication diagram*. It takes the form of the directed graph with nodes representing agents and edges representing communication channels between ports of agents. Alvis communication diagrams allow users to group a set of agents into a subsystem that is represented as a hierarchical agent at the higher level. Hierarchical communication diagrams are used to simplify modelling of more complex systems, but they do not influence the model semantics. Thus we

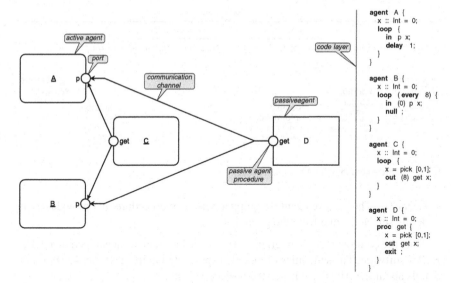

```
agent  A {
  x :: Int = 0;
  loop {
    in  p x;
    delay  1;
  }
}

agent  B {
  x :: Int = 0;
  loop  (every  8) {
    in  (0) p x;
    null ;
  }
}

agent  C {
  x :: Int = 0;
  loop {
    x = pick [0,1];
    out  (8) get x;
  }
}

agent  D {
  x :: Int = 0;
  proc  get {
    x = pick [0,1];
    out  get x;
    exit ;
  }
}
```

**Fig. 1** Example of Alvis model

will consider only flat models in the remainder part of the article. For more informa-
tion on hierarchical communication diagrams see [15]. The second level contains the
Alvis code that defines the behaviour of active and passive agents. The small set of
Alvis statements is supported by the Haskell functional programming language [12].
Haskell is used to define parameters, data types and data manipulation functions.

An example of Alvis model is shown in Fig. 1. The model is composed of three
active agents and one passive agent. Agents *A* and *B* compete for data provided by
agent *C* and for access to the agent *D* procedure. The behaviour of these agents
was defined for illustrating typical Alvis statements suitable for modelling real-time
systems.

- Agent *A*—The agent collects an integer via port *p*. It uses the blocking communication [11] i.e. after initialisation of a communication it waits until agent *C* provides
  the value or procedure *D.get* is accessible. After collecting an integer the agent is
  postponed for 1 time-unit. This behaviour is repeated inside the infinite loop.
- Agent *B*—The agent uses the *loop every* statement. It means that the contents of
  the loop is repeated every 8 time-units. The agent uses the non-blocking communication with argument 0. It means that the statement finalises a communication
  with agent *C* or calls *D.get* procedure that must be accessible. If it is not possible,
  the communication is abandoned. The *null* statement is necessary at the end of a
  *loop every* statement.
- Agent *C*—The agent randomly selects a value from the given list, assigns it to
  parameter *x*, and sends it via port *get*. If the agent initialises a communication it
  waits at most 8 time-units for finalisation. Otherwise, the communication is abandoned. This behaviour is repeated inside the infinite loop.

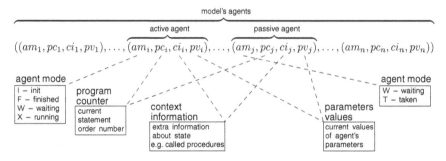

**Fig. 2** Representation of an Alvis model state

- Agent $D$—The passive agent is equipped with one procedure that provides a randomly selected value from the given list.

  Both the communication diagram and the code layer can be developed using *Alvis Editor* software. The semantics of models is presented in the next section. For more details about the Alvis syntax see the project website.[1]

## 3 Model Semantics

A *state of an agent X* is a tuple $S(X) = (am(X), pc(X), ci(X), pv(X))$, where $am(X)$, $pc(X)$, $ci(X)$ and $pv(X)$ denote *agent mode, program counter, context information list* and *parameters values* of the agent $X$ respectively. A *state of an Alvis model* is a sequence of such four-tuples as shown in Fig. 2 [15].

An active agent can be in one of the following modes:

- *Finished* (F)—It means that the agent has finished its work.
- *Init* (I)—This is the default mode for agents that are inactive in the initial state.
- *Running* (X)—It means that the agent is performing one of its statements.
- *Waiting* (W)—It means that the agent is waiting for an event e.g. releasing a currently inaccessible procedure

  A passive agent is in the *waiting* mode if it is inactive and waits for another agent to call one of its accessible procedures or in the *taken* (T) mode if it is executing one of its procedures.

  The program counter points out the current statement of the corresponding agent. The context information list contains additional information about the current state of the corresponding agent e.g. if the agent is in the *waiting* mode, $ci$ contains information about events the agent is waiting for. In case of passive agents in the *waiting* mode, the context information contains list of accessible procedures. The *parameters values tuple* contains the current values of the agent parameters. The initial state for the model from Fig. 1 is as follows:

---

[1]http://alvis.kis.agh.edu.pl.

**Table 1** Alvis transitions ($\alpha^0$ system layer)

| Transition | Arguments | Description |
|---|---|---|
| *TDelay* | *Agent Int* | *Delay* statement execution |
| *TExec* | *Agent Int* | *Exec* statement execution |
| *TExit* | *Agent Int* | *Exit* statement execution |
| *TIn* | *Port Int* | Initialisation of communication, *in* statement |
| *TInAP* | *Port Port Int* | Calling procedure by active agent, *out* statement |
| *TInPP* | *Port Port Int* | Calling procedure by passive agent, *out* statement |
| *TInF* | *Port Port Int* | Finalisation of communication with active agent, *in* statement |
| *TJump* | *Agent Int* | *Jump* statement execution |
| *TLoop* | *Agent Int* | Entering *loop* statement |
| *TLoopEvery* | *Agent Int* | Entering *loop every* statement |
| *TNull* | *Agent Int* | *Null* statement execution |
| *TOut* | *Port Int* | Initialisation of communication, *out* statement |
| *TOutAP* | *Port Port Int* | Calling procedure by active agent, *out* statement |
| *TOutPP* | *Port Port Int* | Calling procedure by passive agent, *out* statement |
| *TOutF* | *Port Port Int* | finalisation of communication with active agent, *out* statement |
| *TSelect* | *Agent Int* | Entering *select* statement |
| *TStart* | *Agent Int* | *Start* statement execution |
| *STInAP* | *Port Port Int* | System version of *TInAP*—for wake up purposes |
| *STInPP* | *Port Port Int* | System version of *TInPP*—for wake up purposes |
| *STOutAP* | *Port Port Int* | System version of *TOutAP*—for wake up purposes |
| *STOutPP* | *Port Port Int* | System version of *TOutPP*—for wake up purposes |
| *STDelayEnd* | *Agent Int* | Termination of agent's suspension |
| *STLoopEnd* | *Agent Int* | Termination of current *loop every* run |
| *STInEnd* | *Port Int* | Abandonment of communication, non-blocking *in* statement |
| *STOutEnd* | *Port Int* | Abandonment of communication, non-blocking *out* statement |
| *STTime* | *Int* | Passage of time |

$$((X, 1, [\ ], 0), (X, 1, [\ ], 0), (X, 1, [\ ], 0), (W, 0, [out(get)], 0)$$

Execution of Alvis statements is described using the *transition* idea. For example execution of the *loop* statement i.e. entering a loop if its guard is true (or there is no guard) is represented by the *TLoop* transition. The activity of a transition is always considered for a given agent and statement. The set of Alvis transitions for models with $\alpha^0$ system layer is given in Table 1.

The transitions *TInAP, TInPP TInF* and *TIn* represent the *in* statement. The use of four transitions for single statement is associated with the variety of situations in

which the statement can be used. Let us focus on the model from Fig. 1. When agent
*A* executes the statement in p x; it means that *A* wants to collect a value via port
*p* and assign it to parameter *x*. The agents does not know whether the value will be
provided by the active agent *C* or the passive agent *D*. On the other hand, from the
*Alvis Compiler* point of view these situations must be distinguished and they are
represented by different transitions. Similarly, four transitions are used to represent
the *out* statement. For more information about different Alvis communication modes
see [11].

Despite the set of transitions that directly represent execution of some statements,
names of these transitions start with capital T, there are transitions that represent
some activities of the model *runtime environment* (so-called *system transitions*)—
names of these transitions start with capital S. These transitions represent waking up
of an agent that, from some reasons, is in the *waiting* mode. There is one exception,
the *STTime* transition, that represents the passage of time. It is used when there are no
transitions available in the current moment (e.g. all active agents are in the *waiting*
mode) but at least one of them will be enabled in some future moment. It is used to
shift the value of the global clock.

Table 1 contains the list of all Alvis transitions for time $\alpha^0$ models. The set of
transitions used in a given model depends on the statements used in the code layer.
For example the following transitions can be enabled for agent *A* from the considered
example: *TLoop A* 1, *TInF A.p C.get* 2, *TInAP A.p D.get* 2, *TIn A.p* 2, *STInAP A.p
D.get* 2, *TDelay A* 3, *STDelayEnd A* 3. It should be stressed that we have four different
transitions for the second statement.

## 3.1 Enable Rules

For each of the transitions from Table 1 we can define *enable* and *firing* rules. Enable
rules define conditions when the given transition is enabled. The firing rules define
how the given transition influences on the change of the current model state. For
active agents and transitions: *TDelay*, *TExec*, *TExit*, *TJump*, *TLoop*, *TLoopEvery*,
*TNull*, *TSelect*, and *TStart* the given transition is *enable* iff the agent is in the *running* mode and the corresponding statement is the current statement, what is indicated by the agent program counter. In case of communication transitions additional
conditions must be fulfilled:

- *TInAP*—There exists an accessible procedure connected with the considered port
  of the active agent and none procedure is executed for the agent currently.
- *TInF*—There exists an active agent that already initialised a communication (*out*
  statement) via a port connected with the considered port of the active agent, that
  executes the transition, and none procedure is executed for the agent currently.
- *TIn*—Transitions *TInAP*, *TInF* are not enabled and no procedure is executed for
  the agent currently—If it is not possible to call a procedure or finalise a communication, the transition initialises a communication and moves the agent to the

*waiting* mode. If it waits for a currently inaccessible procedure then the *STInAP* transition wakes up the agent when the procedure is accessible.

Enable conditions for *TOut*, *TOutAP*, *TOutF*, and *STOutAP* transitions are defined analogously. In case of passive agents the conditions are similar but the given passive agent must be in the *taken* mode and the active agent in which context it works must be in the *running* mode. For passive agents, transitions *TInPP*, *TOutPP*, *STInPP*, and *STOutPP* are used instead of corresponding ...*AP* transitions.

The system transitions *STDelayEnd*, *STInEnd*, *STOutEnd* are enabled if the agent is in the *waiting* mode after executing the corresponding statements (in case of passive agent the agent is in the *taken* mode and the context agent is in the *waiting* mode) and the waiting time has elapsed. Finally, the *STLoopEnd* transition is enabled if the agent has finished executing the contents of the corresponding *loop every* statement and the period of the loop has expired.

## 3.2  Firing Rules

As it is presented in Sect. 4, in case of time models a few transitions can be executed in parallel. However, to describe the *firing rules* we consider the results of executing of individual transitions in a given state $s$. Let $nextpc(n)$ denote the next program counter determined on the basis of the code structure for the considered agent and the current program counter $n$. For example, if we consider a *TLoop* transition then $nextpc(n)$ is equal to the number of the first statement inside the loop if the guard is satisfied (or there is no guard) or the number of the first statement after the loop otherwise. It is assumed that $nextpc(n) = 0$ if there is no next statement.

Assume we consider the result of a transition firing for agent $X$, $n$ denotes the number of the statement the considered transition refers to, and $context(X)$ denotes the active agent in which context $X$ works if $X$ is a passive agent. Firing of the *TExec*, *TJump*, *TLoop*, *TLoopEvery*, *TNull*, *TSelect* or *TStart* transition sets $pc(X) = nextpc(n)$. Moreover, the *TExec* transition updates the value of the parameter used as the left-hand side of the assign operator; *TLoopEvery* transition adds $timer(n, d)$ entry to $ci(X)$, where $d$ represents the number of time-units to the end of the current loop run; and the *TStart* transition set its argument (agent) to the *running* mode and its program counter to 1 if the agent is in the *init* mode. If $X$ is an active agent and $nextpc(n) = 0$ then any of these transitions (except *TJump* and *TLoopEvery*) sets $am(X) = F$ and $ci(X) = [ ]$. The *null* statement is also used to point out the end of the contents of a *loop every* statement. In such a case, the corresponding *TNull* transition sets $am(X) = W$ (or $am(context(X)) = W$ if $X$ is a passive agent), and $pc(X)$ to the number of the corresponding *loop every* statement.

Firing of the *TDelay* transition sets $am(X) = W$ (or $am(context(X)) = W$ if $X$ is a passive agent) and adds $timer(n, d)$ entry to $ci(X)$, where $d$ represents the number of time-units of the suspension.

If $X$ is an active agent then firing of the *TExit* transition sets $pc(X) = 0, am(X) = F$, and $ci(X) = [ ]$. If $X$ is a passive agent then firing of the *TExit* transition ends the current procedure (let us denote it by $X.p$) i.e. sets $am(X) = W$, $pc(X) = 0$, and $ci(X)$ to the set of $X$ procedures accessible in the new state. Moreover, if the procedure has been called by an agent $Y$ then the $proc(X.p)$ entry is removed from $ci(Y)$ and $pc(Y)$ is set to its next value (if it is 0 and $Y$ is an active agent then also $am(Y) = F$, and $ci(Y) = [\ ]$).

Firing of the transition $TInAP$ $X.p$ $Y.q$ $n$ (or $TInPP$, $TOutAP$, $TOutPP$ with the same arguments) inserts $proc(Y.q)$ entry to $ci(X)$ and sets $am(Y) = T$, $ci(Y) = [\ ]$, and $pc(Y)$ to the number of the first statement in $Y.q$ procedure. System transitions $STInAP$ and $STInPP$ additionally changes $am(X)$ ($am(context(X))$ if $X$ is a passive agent) from $W$ to $X$ and removes $in(p)$, $timer(n, d)$ entries from $ci(X)$ (the $timer(n, d)$ entry is used only for non-blocking communication). System transitions $STOutAP$ and $STOutPP$ work similarly but it removes $out(p)$ entry instead of $in(p)$. The result of the $TIn$ $X.p$ $n$ transition firing depends on the type of communication:

1. *Non-blocking with time $d = 0$*: sets $pc(X) = nextpc(n)$ (if it is 0 and $X$ is an active agent then also $am(X) = F$, and $ci(X) = [\ ]$).
2. *Non-blocking with time $d > 0$*: sets $am(X) = W$ ($am(context(X)) = X$ if $X$ is a passive agent), and inserts $in(p)$, $timer(n, d)$ entries into $ci(X)$.
3. *Blocking, $X$ is an active agent*: sets $am(X) = W$ and inserts $in(p)$ entry into $ci(X)$.
4. *Blocking, $X$ is a passive agent, $X.p$ is non-procedure port*: sets $am(context(X)) = W$ and inserts $in(p)$ entry into $ci(X)$.
5. *Blocking, $X$ is passive agent, $X.p$ is procedure port*: sets $pc(X) = nextpc(n)$, and updates value of the corresponding parameter (if a value has been sent).

The $TOut$ transition works similarly but $out(p)$ entry is inserted instead of $in(p)$ and in case of a procedure port there is no parameter update.

Firing of the $TInF$ $X.p$ $Y.q$ $n$ transition updates value of the corresponding parameter of agent $X$ (if a value has been sent), sets $pc(X) = nextpc(n)$, $am(Y) = X$, $pc(Y) = nextpc(m)$ (where $m$ is the current value of $pc(Y)$), and removes $out(q)$ and $timer(n, d)$ entries from $ci(Y)$. If $nextpc(n) = 0$ or $nextpc(m) = 0$ then for the corresponding agent the mode is set to $F$ and the context list to $[\ ]$. The $TOutF$ transition works similarly but $in(q)$ entry is removed instead of $out(q)$ and a parameter of $Y$ agent is potentially updated.

Firing of the $STLoopEnd$ transition sets $am(X) = X$ ($am(context(X)) = X$ if $X$ is a passive agent), and removes $timeout(n)$ entry from $ci(X)$. Firing of a $STDelayEnd$ transition additionally sets $pc(X) = nextpc(n)$ (if it is 0 and $X$ is an active agent then also $am(X) = F$, and $ci(X) = [\ ]$).

Firing of the $STInEnd$ transition sets $am(X) = X$ ($am(context(X)) = X$ if $X$ is a passive agent), removes $in(p)$ and $timeout(n)$ entries from $ci(X)$, and sets $pc(X) = nextpc(n)$ (if it is 0 and $X$ is an active agent then also $am(X) = F$, and $ci(X) = [\ ]$). A $STOutEnd$ works similarly but it removes $out(p)$ entry instead of $in(p)$.

The presented *enable* and *firing* rules show how the firing of a transition affects the changes of states. More precise definitions of the rules but in the context of the Intermediate Haskell Representation (IHR) of Alvis models are presented in the language manual [18]. The next section describes how a new state is determined if we take into account a set of transitions executed in parallel and the passage of time.

# 4   LTS Graphs for Time Models

To verify an Alvis model's properties using model checking techniques [2] it is necessary to generate the model state-space first. We use Labelled Transition Systems (LTS graphs) to represent such state-spaces. Nodes of an LTS graph represent reachable model states. Labels of arcs provide two pieces of information: a set of transitions that are executed in parallel and lead from the corresponding arc source state to the arc destination state and the time that elapsed between these two consecutive states. The initial part of the LTS graph for the model from Fig. 1 is shown in Fig. 3. This section focuses on the most important parts of the LTS generation algorithm implemented in Alvis Toolkit.

The generation of an LTS graph for time models starts with the initial state that has the serial number 0 and is the initial node of the LTS graph. The initial state is the first *current* node. For any current node the set of outgoing arcs and finally direct successors are generated. Let $s$ denote the model state represented by the current node. We start with generation of the set of transitions enabled in state $s$ according to the enable rules presented in Sect. 3. For example the set of transitions enabled in the initial state of the considered model contains three elements: *TLoop A* 1, *TLoopEvery B* 1, *TLoop C* 1. If such a set does not contain any communication transition then all the transitions can be executed in parallel as shown in Fig. 3.

By default duration of each statement (transition) is equal to 1 time-unit. One can define the individual value of duration for each model statement. This is done using the *duration* function implemented in Haskell. In case of the considered model the function is defined as follows:

```
duration :: Agent -> Int -> Int
duration A 1 = 1
duration A 2 = 2
duration A 3 = 1
duration B 1 = 2
duration B 2 = 3
duration B 3 = 0
duration C 1 = 1
duration C 2 = 2
duration C 3 = 3
duration D 1 = 2
duration D 2 = 3
duration D 3 = 1
duration _ _ = 1
```

This means that the three enabled transitions have different durations (1, 2, and 1 respectively) and we cannot move from the initial state directly to a state where all the transitions are finished. In this case, we can move only 1 time-unit forward. Thus, the new state describes a situation when one of the transitions is still under execution (see state 1, Fig. 3). The *sft(n)* (*step finish time*) entry used in *ci(B)* points out the number of time-units necessary to finish the current transition.

The set of transitions executed in parallel is called *multi-step*. The LTS graph generation algorithm determines the maximal *time shift* for each multi-step. This value is selected so as not to lose any information about the changes of states of the analysed system. The algorithm takes into account not only the duration of each

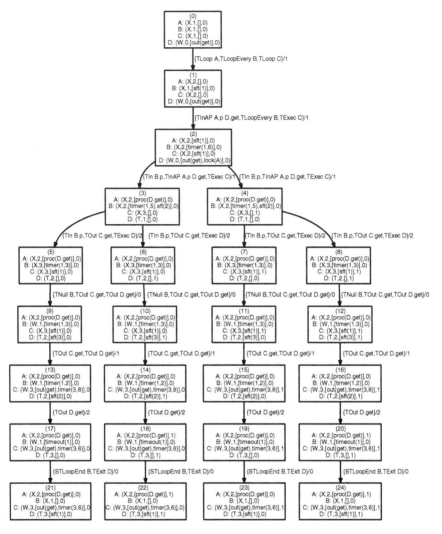

**Fig. 3** Initial part of the LTS graph for the model from Fig. 1

transition in the multi-step but also the arguments of context entries such as *sft* or *timer*.

The result of a multi-step execution consists not only of the effects of single transitions execution but also the results of the time shift i.e. the arguments of all timers and *sft* entries are reduced by the value of the time shift, even if an agent does not execute a transition in the given multi-step. Moreover, if the time argument of a timer is reduced to 0, then the entry is replaced with *timeout* one (e.g. see state 17, Fig. 3).

If the set of enabled transitions contains communication transitions then *conflicts* may arise e.g. two transitions are enabled but they cannot be executed in parallel.

This is the case with the state 437:

$$(X, 2, [\ ], 0), (X, 2, [timer(1, 6)], 0), (X, 3, [sft(1)], 0), (W, 0, [out(get)], 0))$$

The set of enabled transitions contains the following elements: *TInAP A.p D.get* 2, *TInAP B.p D.get* 2, and *TOut C.get* 3. Thus, we have two agents A and B that compete for the access to the same passive agent. Of course these three transitions cannot be executed in parallel. The set of enabled transitions must be divided into two multi-steps.

The transitions *TInAP A.p D.get* 2 and *TInAP B.p D.get* 2 cannot be placed in the same multi-step, but both agents A and B can execute a transition in each of this two multi-steps. If we chose the *TInAP A.p D.get* 2 transition then the port *D.get* is already inaccessible for agent B, but the agent can execute the *TIn B.p* 2 transition. Thus both multi-sets contain three transitions as shown in Fig. 4 (see labels for node 437 outgoing arcs). States 453, 454 and 478 illustrate a situation when agent D is still in the *waiting* mode, because the corresponding *TInAP* is not finished yet, but

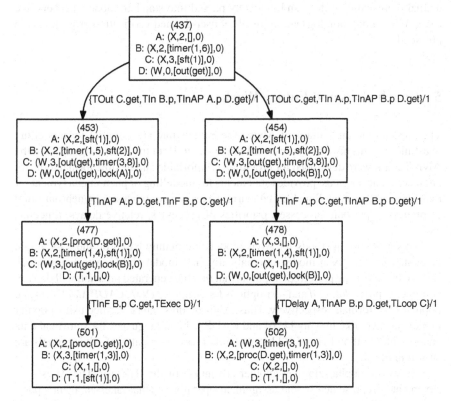

**Fig. 4** Part of the LTS graph for the model from Fig. 1

it is not accessible for other agents, because the *TInAP* is already started. This is indicated by the *lock* entry included into the context information list.

Stages like determining the list of enabled transitions, dividing the set into multi-steps, determining time shift for each multi-step are performed for each reachable state. The final LTS graph contains a node for each reachable state and an arc for each executed multi-step. The LTS graph is generated automatically. The Alvis language is supported by computer tools called *Alvis Toolkit*. Models can be designed with *Alvis Editor* that provides essential editing features, such as: diagram edition, basic tools for alignment and colouring, automatic creation and removal (flattening) of hierarchical pages [15], textual layer addition with syntax colouring and code folding. An Alvis model stored in an XML file is then processed by *Alvis Compiler* that generates the IHR for the model. The generated Haskell file can be modified by the user. For example user-defined verification algorithms, a priority management algorithm, user-defined *duration* function (assignment of duration to each model statement), user-defined *main* function, etc. can be included into the file before compilation. Finally, depending on the used *Alvis Compiler* options and user's optional code modifications, as a result of the Haskell program execution, the LTS graph in various textual representation that can be directly passed into standard model checkers like e.g. *CADP*, simulation logs and/or result of user-defined verification procedures are provided.

## 5 Conclusions and Future Work

The most important features of Alvis time models that are essential for modelling real-time systems have been presented in the paper. Both the Alvis language and the Alvis Toolkit were developed for providing comfortable and flexible formal tools for engineers. The language provides statements for modelling of phenomena typical for real-time systems modelling like: concurrent executing of processes, synchronisation of processes, periodic processes, priorities of processes, relative delays, timeouts, etc.

The verification methods for Alvis models are mainly based on the LTS graph, generated for the given model automatically, and model checking techniques [2]. Users can choose the preferred output format that the generated program will deliver, compiler expose *-dot*, *-ald* and *-csv* options for *Graphviz DOT*, *CADP Aldebaran* and popular *CSV* formats respectively. Thus, Alvis models can be verified using popular model checkers like nuXmv [4, 6] and CADP [7] and languages for statistical data analysis like Python [8] and R [10]. Possible processing paths of Alvis models are shown in Fig. 5.

It is worth emphasizing yet another advantage of the IHR use. The Alvis *exec* statement is represented by the assignment operator (=) that takes an agent's parameter as its left-hand side argument and a Haskell expression as the right-hand side argument. The statement evaluates the expression and assigns its result to the parameter. This is represented as a single step in the LTS graph, but the Haskell

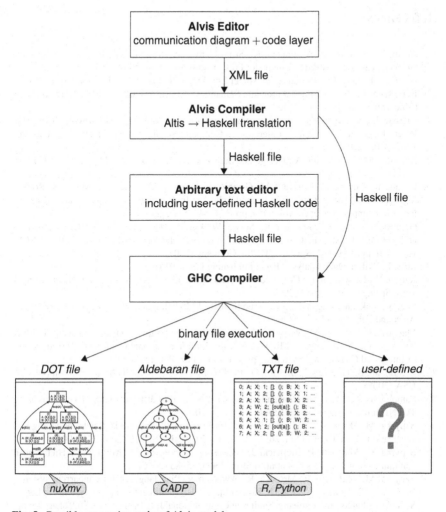

**Fig. 5** Possible processing paths of Alvis models

expression may contain any user-defined function and may represent complex operations on some data. It is an easy way to include, for example, artificial intelligence systems like rule-based systems, decision trees, neural networks, etc. into Alvis models. An example of including a decision support system into an Alvis model of a railway traffic management system is presented in [17].

The presented version of Alvis time models supports only so-called $\alpha^0$ system layer. There is also $\alpha^1$ system layer under development. This layer is based on the assumption that there is only one processor and all active agents compete for access to it. This will allow Alvis to be used for modelling of software for single-processor embedded systems.

# References

1. Aceto, L., Ingófsdóttir, A., Larsen, K., Srba, J.: Reactive Systems: Modelling, Specification and Verification. Cambridge University Press, Cambridge, UK (2007)
2. Baier, C., Katoen, J.P.: Principles of Model Checking. The MIT Press, London, UK (2008)
3. Bengtsson, J., Yi, W.: Timed Automata: Semantics, Algorithms and Tools. Lecture Notes on Concurrency and Petri Nets 3098 (2004)
4. Biernacki, J.: Alvis models of safety critical systems state-base verification with nuXmv. In: Proceedings of the Federated Conference on Computer Science and Information Systems, pp. 1701–1708 (2016)
5. Bozzano, M., Villafiorita, A.: Design and Safety Assessment of Critical Systems. CRC Press (2011)
6. Cavada, R., Cimatti, A., Dorigatti, M., Griggio, A., Mariotti, A., Micheli, A., Mover, S., Roveri, M., Tonetta, S.: The nuXmv symbolic model checker. In: Computer Aided Verification, Lecture Notes in Computer Science, vol. 8559, pp. 334–342. Springer (2014)
7. Garavel, H., Lang, F., Mateescu, R., Serwe, W.: CADP 2006: a toolbox for the construction and analysis of distributed processes. In: Computer Aided Verification (CAV'2007). LNCS, vol. 4590, pp. 158–163. Springer, Berlin, Germany (2007)
8. Idris, I.: Python Data Analysis. Packt Publishing Ltd. (2014)
9. Jensen, K., Kristensen, L.: Coloured Petri Nets. Modelling and Validation of Concurrent Systems. Springer, Heidelberg (2009)
10. Lee, A., Ihaka, R., Triggs, C.: Advanced Statistical Modelling. Course Notes for University of Auckland Paper STATS 330 (2012)
11. Matyasik, P., Szpyrka, M., Wypych, M., Biernacki, J.: Communication between agents in Alvis language. In: Proceedings of Mixdes 2016, the 23nd International Conference Mixed Design of Integrated Circuits and Systems, pp. 448–453. Łódź, Poland (2016)
12. O'Sullivan, B., Goerzen, J., Stewart, D.: Real World Haskell. O'Reilly Media, Sebastopol, CA, USA (2008)
13. Samolej, S., Rak, T.: Simulation and performance analysis of distributed internet systems using TCPNs. Informatica (Slovenia) 33(4), 405–415 (2009)
14. Szpyrka, M., Biernacki, J., Biernacka, A.: Tools and methods for RTCP-nets modelling and verification. Arch. Control Sci. 26(3), 339–365 (2016). doi:10.1515/acsc-2016-0019
15. Szpyrka, M., Matyasik, P., Biernacki, J., Biernacka, A., Wypych, M., Kotulski, L.: Hierarchical communication diagrams. Comput. Inform. 35(1), 55–83 (2016)
16. Szpyrka, M., Matyasik, P., Mrówka, R.: Alvis—modelling language for concurrent systems. In: Bouvry, P., Gonzalez-Velez, H., Kołodziej, J. (eds.) Intelligent Decision Systems in Large-Scale Distributed Environments. Studies in Computational Intelligence, vol. 362, chap. 15, pp. 315–341. Springer (2011)
17. Szpyrka, M., Matyasik, P., Podolski, L., Wypych, M.: Simulation of multi-agent systems with Alvis toolkit. In: Proceedings of ICAISC 2017, the 16th International Conference on Artificial Intelligence and Soft Computing, LNAI, vol. 10246. Springer (2017). doi:10.1007/978-3-319-59060-8_54
18. Szpyrka, M., Matyasik, P., Wypych, M., Biernacki, J., Podolski, L.: Alvis Modelling Language (2017). http://alvis.kis.agh.edu.pl

# Control Operation Flow for Mobile Access Control with the Use of MABAC Model

**Aneta Majchrzycka and Aneta Poniszewska-Maranda**

**Abstract** As mobile devices tend to become more and more essential part of every-day life the significance of security of applications designed for the use on such devices is constantly growing. Most of the mobile platform providers assure sophisticated security mechanisms which aim to prevent threatening applications to be published as well as they make certain security features available to use. Nevertheless, the design and implementation of these features leaves a vast area where potential enhancements can be brought to life and mistakes avoided. In order to address this field of mobile application development MABAC, an access control model was introduced which aims to characterize the access control flow specific for mobile systems. The paper describes the crucial processes connected with the MABAC model and brings closer the details of the control flow in the mobile application designed according to the rules of the model.

## 1 Introduction

Mobile applications have become more and more popular among mobile device users which increases the requirement for fast and reliable development of such applications. Not only stems this trend from the changing everyday lifestyle of people but also from the changing tendencies in corporate management which aim to use the newest technological advances in order to accelerate business process within the companies. These observed alterations in the way people want to make use of technology promise a growing number of bright possibilities which lie ahead, nevertheless one should not forget about the increased vulnerability of the confidential data which is being exchanged in consequence. Thus, the even more urging need for assuring proper security of data emerges [1–3].

Virtually all of the mobile applications which are installed on our smartphones and tablets are parts of larger, more complex systems which both to provide us

A. Majchrzycka · A. Poniszewska-Maranda (✉)
Institute of Information Technology, Lodz University of Technology, Łódź, Poland
e-mail: aneta.poniszewska-maranda@p.lodz.pl

© Springer International Publishing AG 2018
P. Kosiuczenko and L. Madeyski (eds.), *Towards a Synergistic Combination of Research and Practice in Software Engineering*, Studies in Computational Intelligence 733, DOI 10.1007/978-3-319-65208-5_13

with certain functionalities as well as gather information about the application users. Many times, unknowingly, we share sensitive data like email addresses, phone numbers, device identifiers which are then stored by application providers and can be used mainly for personalization and marketing purposes. By using the specific mobile application, we, willingly or not, agree to such terms which in most cases will be harmless unless our device or user profile becomes a target of the attack which aims to capture the sensitive data [4–6].

This is the reason why significant amount of time and effort should be placed to create an application in such a way so that the data and control flow within the application was implemented in the most secure and controlled manner possible. There exist numerous guidelines for developers specifying how to design the mobile application and which security standards should they conform to, nevertheless there is no strict model or policy which would enforce the usage of certain security mechanisms which in most of the cases seems vital [7, 8].

In order to address this field of mobile application development the MABAC, an access control model was introduced which aims to characterize the access control flow specific for mobile systems. The problem described in the paper concerns the crucial processes connected with the MABAC model and brings closer the details of the security control flow in the mobile application designed according to the rules of the model.

The presented paper is structured as follows: Sect. 2 gives the outline of Secure Development Strategy (SDS) created for development of mobile application security. Section 3 deals with access control model approach with SDS. Section 4 presents the access control approach based on mobile applications—mARBAC model while Sect. 5 provides the description of process connected with the data and control flow for mobile access control with the use of MABAC model.

## 2   Secure Development Strategy for Mobile Application Development

A suggested strategy for secure mobile application development was presented in [9] as Secure Development Strategy (SDS). It revolves around the idea of a threefold security assurance and divides the data management into three areas:

- storage,
- access,
- transfer.

out of which all should be taken care of in terms of security in order to assure minimal level of application safety. The strategy enumerates a list of mechanisms which constitute basic security features in each of the areas.

The first pillar of the SDS storage concerns solely the client-side of the system i.e. the mobile application. The major assumptions of data storage pattern embrace sensitive data encryption, limitation and restricted access. The second pillar of SDS strategy concerns the access to data. This comes from the fact that mobile applications need to communicate with external services and other applications. The major assumptions of this area of security embraces three mechanisms which aim to enable identification of the user requesting access to application resources.

Data transfer pattern refers to all mechanisms which involve the exchange of data between the mobile application and the external services. These mechanisms should incorporate in their action flow the additional security procedures—data encryption, the use of security keys and the verification of the requests integrity. They form the last third pillar of SDS strategy [9, 10].

All the layers of data management distinguished by the SDS overlap one another and the access area can be viewed as the linking component between the two others. Thus, its importance can be noted as twofold in this case and further development of this area was pursued. It consisted mainly in adjusting an access control model based on roles (RBAC [11, 12]) to meet the characteristic features of the mobile applications and systems. It led to the creation of the mobile Application-Based Access Control model (MABAC) described in more details in the next section.

# 3 Access Control Model Approach with SDS

The SDS strategy distinguishes three areas of data management in terms of data security in a mobile system. These areas are strictly connected with the architecture of a mobile system which simplified is presented in Fig. 1 in the upper part of the schema. The basic components of the mobile system embrace not only the mobile application installed on the mobile device but also a server application (API) and database residing on the server-side of the system. Other visible components:

- administration module and
- alerting module

are additions to the SDS strategy assumptions and aim to control the traffic of requests incoming to and outgoing from the server. They can contribute to increasing security strictly in terms of the network traffic and are presented here to allow the general overview of the mobile system architecture assumed by the SDS strategy.

The implementation of the ready-to-use classes and methods which would assure security of any mobile applications seems to suit the current needs for simultaneous time-efficiency and safety. That is why the prototype framework *iSec* was developed. Its main components correspond to three pillars of the security model presented in the previous section: storage, access and transfer. Additionally, addressing a practical need for a fast login mechanism, the component generating classes and controllers indispensable while creating login views was designed and incorporated as a part of this framework. This mechanism enables also the use of roles, which are frequently applied for the mobile applications.

As far as the access area security is concerned, all mechanisms introduced by the SDS are designed to be used on the mobile device. There are three main defined

**Fig. 1** Mobile system
architecture assumed by SDS
strategy

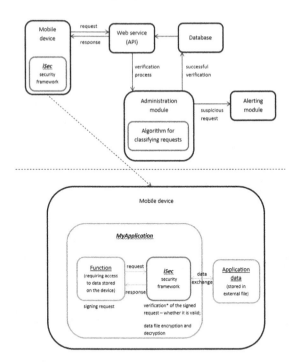

mechanisms and they are recommended for implementation in order to increase the
application security. These mechanisms include:

- *Geolocation*—mechanism known widely, its various forms are applied by leading
  companies like Google and banking industry solutions. The aim of geolocation in
  terms of mobile systems is to identify whether request i.e. access to data stored
  on the server or device is being made from a significantly remote location which
  would indicate the possibility of an attempt to steal the data. Geolocation is a
  means of detecting the potential threats to the system and it does not give one
  hundred percent certainty, however it can vitally contribute to the overall application safety [6, 13].
- *Application-device identifier (ADID)*—it is a common practice to send the unique
  device identifier (UDID) across the network as a means of identifying the specific device within the system. It however is often used by the companies to direct
  marketing campaigns towards specific devices and in some aspects, violates the
  privacy of mobile phone users as a sensitive data piece like UDID is leaked (often
  unnecessarily). That is why a custom identifier is suggested as an alternative.
  ADID is a unique combination of the partial UDID identifier and the timestamp
  indicating the date and time of the installation of the application. Thanks to this
  the uniqueness can be maintained and the unnecessary data is not spread outside
  the secure scope [4, 7, 8].

- *Access control*—this mechanism is defined by the model for access control designed especially for the purposes of mobile system [9, 11, 13]. The simplified flow and placement of this mechanism with the general application flow is depicted in Fig. 1 in the bottom part of the scheme. The more detailed description of the access control model can be found further in this and the next section.

# 4 Mobile Application Based Access Control Approach

The development of access control policies and models has a long history. It is possibly to distinguish two main approaches. The first one represents the group of traditional access control models, such as Discretionary Access Control (DAC) model [14], Mandatory Access Control (MAC) [14], Role-Based Access Control (RBAC) model [11, 15].

Second approach of access control models corresponds to the temporal models that introduce the temporal features into traditional access control, such as Temporal Authorization model [16], Temporal-RBAC (TRBAC) [17], Temporal Data Authorization model (TDAM) [18] or Generalized Temporal RBAC (GTRBAC) [19].

Currently, traditional access control models, such as Discretionary Access Control (DAC) model [14], Mandatory Access Control (MAC) model [14], Role-Based Access Control (RBAC) model [11] or even Usage Control model [20], are not sufficient and adequate in many cases for information systems, especially modern, dynamic, distributed information systems, which connect different environments by the network. The same situation exist in the aspect of mobile applications and mobile platforms. It caused the creation of new access control model for mobile applications that can encompass the traditional access control ideas and solutions and allow to define the rules for mobile applications and systems, containing both static and dynamic access control aspects. Therefore, to ensure the functionality of the second pillar of SDS strategy, i.e. data access security model, the new access control model approach for mobile applications was proposed.

Actual applications and information systems can contain or work with many different components, applications, located in different places in a city, in a country or on the globe. Each of such components can store the information, can make this information available to other components or to different users. The authorized users accessing the information can change this information, its status, role or other attributes at any time. These changes can cause the necessity of modifications in security properties of accessed data at access control level. Such modifications are dynamic and often should be realized ad hoc because other users from other locations can request the access to the information almost at the same time. Many different users from different locations can access information system. Sometimes, they need to have the direct and rapid access to actual information. However, the conditions of such access are very often dynamic, they can change for example because of actions of other users. It is necessary to ensure the ad hoc checking of current rights of an user basing on their current conditions that can change dynamically.

Therefore, there is the need to have the access control approach that will describe the organization of mobile application/system that should be secured, their structure in proper and complete way, and on the other hand it will be appropriate and sufficient for dynamic application/system.

The core part of formulated mobile access control approach, i.e. *Mobile Application-based Access Control (MABAC) model*, essentially represents the basic elements of access control, such as *subjects*, *objects* and *methods* [11, 20, 21]. We distinguished two main types of subjects in MABAC: user (*User*) and mobile device (*Device*). These two elements are represented by the **Subject** that is the superclass of User and Device. *Subjects* hold and execute indirectly certain rights on the objects. Mobile applications are working on behalf of users who execute them on devices directly or indirectly [9].

Both user, mobile device and subject are identified by their name—a user by *uName*, a devices by *dName* or its identification number *dID* and subject by *sName*:

$$\forall u_i \in U, u_i = (uName_i)$$
$$U - set\ of\ all\ system's\ users, i \in N$$

$$\forall d_i \in D, d_i = (dName_i \mid dID_i)$$
$$D - set\ of\ all\ system's\ devices, i \in N$$

$$\forall s_i \in S, s_i = (sName_i)$$
$$S - set\ of\ all\ system's\ subjects, i \in N$$

Subject permits to formalize the assignment of users and mobile devices to different functions. It is presented as an abstract type, so it can not have direct instances—each subject is either a users or a devices. A mobile application **Application** is a mobile program executed on a device by a user. It needs to obtain an access to the desired data in order to realize the tasks asked by a user, so it represents the system entity, that can obtain some access rights in a system.

The **Session** element represents the period of time during which the user is logged in a system and can execute its access rights or represents the period of time during which the application is working on the device and can also execute its access rights. In our model the *Session* is assigned to the *Subject*, i.e. a user is login on the device during a single session. On the other hand a session is connected with the *application* and this association represents the application that can be activated during one session. Moreover, a session is connected with the *functions* and this association represents the functions that can be activated during one session [10].

The association relation between the subjects and applications is described by the association class **SubjectAttributes** that represents the additional subject attributes (i.e. subject properties) as in usage control. *Subject attributes* provide additional properties, describing the subjects, that can be used in taking decision about granting or revoking the subject an access to certain object—especially *Location* and *Device ID*, but also for example an identity, role, credit, membership.

*Location, L* is expressed by IP address of the device. It is used to determine whether the request is valid (or should be handled) by checking the privileges pursuant to incoming IP address.

Each application allows the user to perform specific tasks related to business process. The application can contain many features that a user can apply. As a consequence, the application can be perceived as a set of functions that this application perform to accomplish a specific job. It is defined by its name *aName* and by the set of associated *aFun* functions:

$$\forall a_i \in A, a_i = \left(aName_i, aFun_i\right)$$

$$A - set\ of\ all\ system's\ applications, i \in N$$

The session is defined by subject's name *sName*, application's name *aName* and by the set of associated *seFun* functions:

$$\forall se_i \in SE, se_i = \left(sName_i, aName_i, seFun_i\right)$$

$$SE - set\ of\ system's\ sessions, i \in N$$

A **Function** is a job-activity within the application with some associated semantics regarding the authority and responsibility conferred on this application. The function can represent a competency to do a specific task-activity and it can embody the authority. The applications are assigned to the function, based on their functionality. The application can have different functions in different cases/situations. It is also possible to define the hierarchy of functions, represented by aggregation relation *FunctionHierarchy*, which represents also the inheritance relations between the functions. The function of the part end of the relation inherits all privileges of parent function [10].

The association relation between applications and functions is described by the association class **ApplicationAttributes** that represents the additional applications attributes (i.e. application properties) that can be used in taking decision about granting or revoking the application an access to an object. *Application attributes* provide additional properties, describing the applications, especially *Security Level* and *Application-Device ID*.

*Security level, Sl* is the ability to set different levels of security for different methods/objects or at the level of entire application. It is necessary therefore, that not all applications require such extensive control and security. There are three levels of security: Ignore, Warn, Block. It is used especially when detected no permission for the request.

*Application-Device IDentifier, ADID* is combined string of unique application ID (the same for all devices using the application, 16 characters, alpha-numeric) and device UDID. Application ID is stored in the configuration file of the server and the configuration of application (*XML config file*). UDID is transmitted during the first application use and stored in the database (encrypted).

Each function can perform one or more operations, so it needs to be associated with a set of related permissions **Permission**. A function can be defined as a set or a sequence (depending on particular situation) of permissions. The access to required object is needed to perform an operation, so necessary permissions should be assigned to corresponding function. Therefore, all the tasks and required permissions are identified and they can be assigned to the application to give it the possibility to perform the responsibilities involved when it realize its functionality. Due to the cardinality constraints, each permission must be assigned to at least one function to ensure the coherence of the whole access control schema [10].

The function is thus a set of permissions and is defined by its name *fName* and by a set of permissions assigned to it *fPerm*:

$$\forall f_i \in F, f_i = \left( fName_i, fPerm_i \right)$$

$$F - set\ of\ all\ system's\ functions, i \in N$$

The permission determines the execution right for a particular method on the particular object. In order to access the data, stored in an object, a message has to be sent to this object. This message causes an execution of particular method **Method** on this object **Object**. Very often the constraints have to be defined in assignment process of permissions to the objects. Such constraints are represented by the authorizations and also by the obligations and/or conditions. Therefore, the *permission* can be presented as a function

$$p(o, m, Cst) \tag{1}$$

where *o* is an object, *m* is a method which can be executed on this object and *Cst* is a set of constraints which determine this permission.

Or using the method's name *pMethod* and object's name *pObject* as follows:

$$\forall p_i \in P, p_i = \left( pMethod_i, pObject_i, \{A, B, C\} \right)$$

$$P - set\ of\ all\ system's\ permissions, i \in N$$

Taking into consideration the subjects attributes (Location and DeviceID) *SubjectAttr = {L, DID}* and application attributes (Security level and ADID) *AppAttr = {Sl, ADID}*, the *permission* can be presented as a function *p(o, m, SubjectAttr, AppAttr, Cst)* or more precisely as:

$$p(o, m, \{L, DID\}, \{Sl, ADID\}, Cst) \tag{2}$$

or as:

$$\forall p_i \in P, p_i = \left( pMethod_i, pObject_i, \{L, DID\}, \{Sl, ADID\}, Cst \right)$$

$$P - set\ of\ all\ system's\ permissions, i \in N$$

**Authorization (A)** is a logical predicate attached to a permission that determines the permission validity depending on the access rules, object attributes and subject attributes. **Obligation (B)** is a functional predicate that verifies the mandatory requirements, i.e. a function that a user has to perform before or during an access. They are defined for the permissions but concerning also the subjects—*Subject* can be associated with the obligations which represent different access control predicates that describe the mandatory requirements performed by a subject before (*pre*) or during (*ongoing*) an access. **Conditions (C)** evaluate the current environmental or application status for the usage decision concerning the permission constraint. They are defined also for the permissions but they concern the session—*Session* can be connected with the set of conditions that represent the features of a system or application [10, 20].

A constraint determines that some permission is valid only for a part of the object instances. Taking into consideration a concept of authorization, obligation and condition, the set of constraints can take the following form $Cst = \{A, B, C\}$ and the permission can be presented as a function:

$$p(o, m, \{L, DID\}, \{Sl, ADID\}, \{A, B, C\}) \tag{3}$$

or as:

$$\forall p_i \in P, p_i = \left(pMethod_i, pObject_i, \{L, DID\}, \{Sl, ADID\}, \{A, B, C\}\right)$$
$$P - set\ of\ all\ system's\ permissions, i \in N$$

According to this, the permission is given to all instances of the object class except the contrary specification.

The **objects** are the entities that can be accessed or used by the applications. The objects can be either privacy sensitive or privacy non-sensitive. The relation between objects and their permissions are additionally described by association class **ObjectAttributes** that represents the additional object attributes (i.e. object properties) that can not be specified in the object's class and they can be used for usage decision process. The examples of object attributes are security labels, ownerships or security classes. They can be also mutable or immutable as subject attributes do.

The **constraints** can be defined for each main element of the model presented above (i.e. subject, session, application, function, permission, object and method), and also for the relationships between the elements. The concept of constraints was described widely in the literature [12, 21]. It is possible to distinguish different types of constraints, static and dynamic, that can be attached to different model elements.

Detailed view of presented Mobile Application-based Access Control (MABAC) model with the set of all elements and relationships is given in Fig. 2.

The example illustrating the presented approach represents the applications of typical university information system. One of such applications can be a mobile application that give an access to some functionalities for system users who could be for example a user "Professor". This user as a *Teacher* can perform some teaching

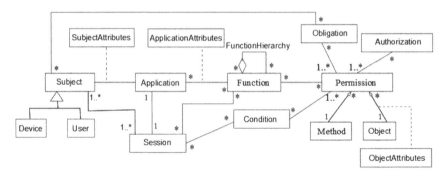

**Fig. 2** Meta-model of MABAC model [10]

activities. Therefore, by an application "Teacher" he has a number of functions, for example share lecture presentation, record exam results, modify results, record final results, etc. Next, the sets of permissions have to be mapped to these functions to grant the access to perform the works required by each function.

In order to describe the particular user of an application the following elements of MABAC model can be determined:

> *subject s* := *Professor*
> *user u* := {*Prof. John Smith*}
> *SubjectAttribute*(*subject s*) := {*position of "Professor" at the department*}
> *Application a* := {*Teacher Application, TeacherApplicationFunctions*}
> *TeacherApplicationFunctions* := {*share lecture presentation, record exam results, modify results, record final results*}
> *listMethods* := *Methods*(*Permissions p*) := {*active*(), *content*(), *chose*(), *validTeacher*(), *getLecture*(*Teacher*), *setExam*(*Lecture, Teacher*), *setGrade*(*Student, Lecture, Exam*), *setValue*(), ...}
> *listObject* := *Objects*(*Permissions p*) := { : *listStudents*, : *listLectures*, : *listExams*, : *Exam*, : *Grade*, ...}
> *listPermissions* := *Permissions*(*Function f*) := {(*content*(), : *listStudents*), (*getLecture*(*Teacher*), : *listLecture*), (*setExam*(*Lecture, Teacher*), : *listExam*), (*setGrade*(*Student, Lecture, Exam*), : *Exam*), ...}
> *ObjectAttribute*(*Object o*) := {*weight of particular grades for the total grade*}
> *listAuthorizations* := *Authorization*(*Permissions p*)
> := {*"Teacher" can visualize only his own Lectures*}
> *listObligations* := *Obligation*(*Permissions p*) := {*permission "(setGrade(Student, Lecture, Exam)" can be executed only after execution of permission "(setExam(Lecture, Teacher), : listExam)"*}
> *listConditions* := *Condition*(*Permissions p*) := {*role "Professor" can realize the permission "(setExam(Lecture, Teacher), : listExam)" only during the business hours*}

The presented example shows the possibilities of developed MABAC approach in domain of modeling and design of an access control for mobile applications.

## 5   Security Control Flow in Mobile Application Based on MABAC Model

As mentioned before the linking component between the areas of the Secure Development Strategy is the access layer. It incorporates three security mechanisms among them the access control model. This model defines the way the functions within the application itself can obtain the rights to perform certain methods on certain data objects.

In order to proceed with the further description of the process connected with data access imposed by the MABAC model it is necessary to define the meaning of certain commonly used notions in the context of the following model.

We refer to an *action* when talking about a single activity performed by the application end user on the interface object, e.g. button or textbox. The action can have a form of a button click or text input activity and it results in firing an *event* which is assigned to the certain interface object. The event is responsible for performing defined *operation* or set of operations which enable the application to go through from one state to another. One operation can be defined by a method or a set of methods. A *method* is a specific implementation of certain functionality within a class and consists of a set of sentences ordered in a sequential manner.

The process of achieving the access to data starts when the end user intends to use the application and performs certain action A. It is depicted in Fig. 3 in a simplified manner.

The action performed by the user enables the sequence of operations executed within the application. The event responsible for handling action A is fired and defined operations are performed. Every operation invokes the call to certain func-

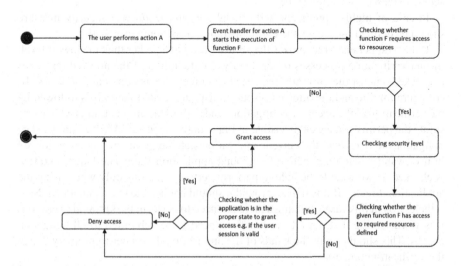

**Fig. 3**   Security control flow based on MABAC for mobile application development

tions, but the functions are not called directly. Instead these calls are being accomplished with the use of the wrapper class which beside calling a function also performs necessary verifications of function privileges.

The wrapper class communicates with the class responsible for parsing XML configuration document. In this document, the privileges of every function are stored. These privileges are defined by the application developer at the stage of the project implementation. One may say that they constitute a separate language of permissions where each entry contains:

- a function name,
- list of allowed resources like keychain keys, local database tables and local files,
- security level which defines the default behaviour of the application during the attempt to execute the function.

The wrapper class first verifies the security level for the given function. The available values of the security level are:

- NoVerificationLevel,
- LowLevel,
- HighLevel.

The *NoVerificationLevel* means that the function probably does not require access to any confidential data and it can be executed directly. If the required function has this value the specified access to the resource and execution ensues immediately. In case of *LowLevel* security the definition access to the resources is granted, but the proper warnings are sent to the user and administrator of the system. *LowLevel* security is an option where the user can use the application freely and is not blocked by internal inconsistencies, however it should only be used with the function requiring access to data of own importance.

In case of highly significant and crucial data the *HighLevel* security measure should be defined as it locks the entire operation from being executed.

If the wrapper class recognizes the security level to be other than the lowest level, further verification processes occur. Firstly, the definition of the allowed resources is checked against the currently requested resource. This is done base on the XML configuration file and the allowed resources list enumerated there. It is followed by the verification of the necessary obligations and authorization constraints which constitute the permission check as visible in Fig. 2 in the depicted MABAC model. The obligations may refer to the current application state being mainly the valid logged in user, valid session and active foreground application state. The latter condition is checked as we want to be able to perform certain functions only when using the application directly. If it is in background it should not be able to execute them. Nevertheless, as in this case sometimes there is a need to perform background operations it is worth noting that not all the functions need to be submitted to all obligation checks. This should lie in the hands of the application developer to properly define the verification process.

All of the verification steps, in case of failure, result in behaviour defined by the set security level of the specific function, i.e. the denial of access or granting access with warnings. In the case, all requirements and conditions are met the access to the resources is granted with no restrictions and the function can be executed without further interference.

The process of statement execution of MABAC access control model allows to handle the flow of data in a controlled manner where all functions requiring access to data, especially sensitive data, are verified methods. In case of capturing the control flow by the malicious software and attempts of function execution when the program is in the background state the verification mechanisms are capable of preventing it if only the MABAC access control model assumptions were implemented within the application control flow.

# 6 Conclusions

Based on the analysis of typical data access solutions in mobile applications and existing access control models, a model dedicated to mobile systems has been developed, taking into account elements that are absent from other models of this type. This model, called mobile application-based access control (MABAC), derives from the traditional RBAC model, taking into account the principles introduced by the SDS model.

The presented Secure Development Strategy for mobile applications introduces three pillars which should be taken into consideration while designing and implementing the mobile applications and their security aspects. No strict standards are provided, rather suggestions for the use of technologies and solutions. The most vital feature of SDS is that it embraces all crucial aspects of mobile application development and systematizes them by defining the concepts and categorization.

The MABAC model allows to define the access control policy based on access request, as traditional access control models, and the access decision can be evaluated while the access to information to which we want to control the usage. All the elements of MABAC approach form fairly complex model to present the features of mobile applications/systems at the access control level. On the other hand, it expresses authorizations, interdictions and obligations that have to be fulfilled in order to obtain the access to dynamic applications/information systems.

Security control flow in mobile application is based on MABAC model and is given for one of the pillars of SDS—access pillar. The practical implementation of MABAC model and components of SDS strategy give the possibility to include the data access control mechanisms in development of mobile applications. The further development will be done to provide an easy to use way for developers to securely store indispensable data directly on the device.

# References

1. Porter Felt A., Finifter M., Chin E., Hanna S., Wagner D.: A survey of mobile malware in the wild. In: Proceedings of 1st ACM Workshop on Security and Privacy in Smartphones and Mobile Devices, pp. 3–14 (2011)
2. Zhou, Y., Jiang, X.: Dissecting android malware: characterization and evolution. In: Proceedings of 33rd IEEE Symposium on Security and Privacy (2012)
3. Michalska, A., Poniszewska-Maranda, A.: Security risks and their prevention capabilities in mobile application development. Inf. Syst. Manage. **4**(2), 123–134 (2015)
4. Apple, iOS Security. http://images.apple.com/ipad/business/docs/iOS_Security_Feb14.pdf (2014)
5. Benedict, C.: Under the Hood: Reversing Android Applications (2001). Infosec Institute (2012)
6. Enck, W., Octeau, D., McDaniel, P., Chaudhuri, S.: A Study of android application security. In: Proceedings of 20th USENIX Security Symposium (2011)
7. Souppaya M. P., Scarfone K. A.: Guidelines for Managing the Security of Mobile Devices in the Enterprise. NIST (2013)
8. Fitzgerald, W.M., Neville, U., Foley, S.N.: MASON: mobile autonomic security for network access controls. Inf. Secur. Appl. **18**(1), 14–29 (2013)
9. Michalska, A., Poniszewska-Maranda, A.: Secure development model for mobile applications. Bull. Polish Acad. Sci. Tech. Sci. **64**(3), 495–503 (2016)
10. Poniszewska-Maranda, A., Majchrzycka, A.: Access control approach in development of mobile applications. In: Younas, M., et al. (eds.) Mobile Web and Intelligent Information Systems, LNCS 9847, pp. 149–162 (2016)
11. Ferraiolo, D., Sandhu, R.S., Gavrila, S., Kuhn, D.R., Chandramouli, R.: Proposed NIST Role-Based Access Control. ACM TISSEC (2001)
12. Ahn, G.-J., Sandhu, R.S.: Role-based authorization constraints specification. ACM Trans. Inf. Syst. Secur. (2000)
13. Alhamed, M., Amir, K., Omari, M., Le, W.: Comparing privacy control methods for smartphone platforms. Eng. Mobile-Enabled Syst. (2013)
14. Castaro S., Fugini M., Martella G., Samarati P.: Database Security. Addison-Wesley (1994)
15. Sandhu, R.S., Coyne, E.J., Feinstein, H.L., Youman, C.E.: Role-based access control models. IEEE Comput. **29**(2), 38–47 (1996)
16. Bertino E., Bettini C., Samarati P.: Temporal access control mechanism for database systems. IEEE Trans. Knowl. Data Eng. **8** (1996)
17. Bertino, E., Bonatti, P., Ferrari, E.: A temporal role-based access control model. ACM Trans. Inf. Syst. Secur. **4**(3), 191–233 (2001)
18. Gal, V., Atluri, V.: An authorization model for temporal data. ACM Trans. Inf. Syst. Secur. **5**(1) (2002)
19. James, B., Joshi, E., Bertino, U., Latif, A., Ghafoo, A.: A generalized temporal role-based access control model. IEEE Trans. Knowl. Eng. **17**(1), 4–23 (2005)
20. Park, J., Sandhu, R.S.: The UCONABC usage control model. ACM Trans. Inf. Syst. Security **7**(1), 128–174 (2004)
21. Poniszewska-Maranda, A.: Modeling and design of role engineering in development of access control for dynamic information systems. Bull. Polish Acad. Sci. Tech. Sci. **61**(3), 569–580 (2013)

# Software Development for Modeling and Simulation of Computer Networks: Complex Systems Approach

Andrzej Paszkiewicz and Marek Bolanowski

**Abstract** The process of creating software for modeling and simulation of phenomena taking place in computer networks should take into account many aspects of their functioning. However, until now, the creation of such software was based on a reductionist approach. This approach is typical for simple or most complicated systems. In contrast, software for modeling and simulation of computer networks should be treated as a complex system. Therefore, the process of its creation should take into account such properties of complex systems as: feedback loop, non-extensivity, indeterminacy, self-similarity, non-additivity, etc. The authors use computer simulators in their work on everyday basis. However, they have generally an outdated, static architecture that prevents their easy and continuous development. Therefore, the authors started working on developing their own model of creating such software and this paper is an introduction to this issue. The authors focused on the selected features of complex systems in the context of the software development process. Based on the feedback loop, a new spiral of software development and modeling for computer networks is presented. The paper also defines the notion of process and functional non-additivity and its importance in the software development process. The presented approach allows for flexible development of the software under consideration in terms of their functionality. The authors also presented examples of application of complex system properties when creating selected functional modules of software for modeling and simulation of computer networks.

**Keywords** Complex systems · Computer networks · Feedback loop · Software development

A. Paszkiewicz (✉) · M. Bolanowski
Department of Complex Systems, Rzeszow University of Technology, Rzeszów, Poland
e-mail: andrzejp@prz.edu.pl

M. Bolanowski
e-mail: marekb@prz.edu.pl

© Springer International Publishing AG 2018
P. Kosiuczenko and L. Madeyski (eds.), *Towards a Synergistic Combination of Research and Practice in Software Engineering*, Studies in Computational Intelligence 733, DOI 10.1007/978-3-319-65208-5_14

# 1 Introduction

The development of computer systems is closely linked to the development of modern software. On the one hand, software should support designing such systems, on the other enable their efficient use. One of important areas connected with computer systems functioning are computer networks. They are basic part of infrastructure for information exchange in such systems as e.g. industrial, military, utility, entertainment, financial ones. Recently, the trend has been observed to connect almost all devices to computer networks (Internet of Things, Industry 4.0). Such activities require taking into account many factors in the process of designing of these networks in order to ensure appropriate quality parameters of the proposed infrastructure. Increasing mobility of end users, dynamically changing requirements as to the value of transmission quality parameters and an increase in the demand for a guarantee of the network reliability should be also kept in mind. In order to meet these challenges, it is necessary to create dedicated software to model and simulate computer networks as well. The scope of its use may be very wide from the conceptual stages, through the implementation of relevant projects, to simulation of the action in specific conditions. It should be also remembered that such software should take into account different levels of the computer network operation i.e. from the physical layer associated with allocation of individual components and their connection, through the logical layer that allows for ensuring routing mechanisms, as well as the application level to simulate real data transmission (generating network traffic with a characteristic flows). Only such functional range allows to create valuable software.

Due to the complexity of such design task, the principle of division into smaller subtasks and the creation of dedicated applications for them is usually used e.g. in terms of resource allocation, network traffic generation, etc. This approach is characteristic for simple systems where correlations between individual elements and phenomena related to them are not significant and they are frequently neglected. So far, the software development process has been based on a reductionist approach typical of simple systems, where well-known laws and existing models can be applied. Often, more complex structures are perceived as complex systems consisting of multiple components. This approach also applies to large scale systems. However, these elements still carry out specific tasks defined and described by simple relationships. Complex systems, however, differ radically from simple systems. They are an emergent structure consisting of elements interacting with one another, and the processes occurring within them range from short-term to long-term ones. Thus, it should be clearly stated that complex systems are not equivalent to the concepts of: large system, distributed system, large computational system, etc. Detailed approach to this class of systems has been described i.e. in the paper [1]. The application of a complex system theory approach enables a better understanding of the phenomena that occur in real-world applications and computer systems.

There are also extensive applications allowing for comprehensive design and simulation of computer networks. Examples of such solutions are Riverbed Modeler [2] or OMNeT++ [3], but they also use models and mechanisms that do not quite correspond to the reality such as traffic generation mechanisms. It should be emphasized that these programs are not designed to automatically search for solutions (Network Design Problem). Dedicated, specialized applications for modeling and optimization of computer networks are created for this purpose.

The software development process, like most design processes, is characterized by large areas of uncertainty and indeterminacy. Of course, the highest level of indeterminacy occurs at the beginning of the software development process, and then is reduced. The paper highlighted the importance of this fact as one of the properties of complex systems. However, the methods and means to account for indeterminacy in the software development process go beyond the scope of this paper.

The main purpose of the paper was to determine the possibilities of using complex system theory to change the approach to software development methodology for simulation of computer networks in relation to the classical stage independent model which is currently used. Thus, the approach proposed in this paper allows to take into account such features and characteristics of complex systems as feedback loop, indeterminacy, non-additivity, etc. [4].

## 2  Classical Approach

Creating software with a high degree of complexity and interrelation of parameters is a complicated task. Software for modeling and computer network simulation belongs to a group of such applications. Therefore, the main task is often divided into smaller sub-tasks in order to simplify the development of a such software. Such an approach results both from generally accepted software development methods (such as SCRUM) and well-known decomposition algorithms, which have been described in numerous publications [5, 6]. The first aspect of decomposition is the division of the software development process into its basic stages such as analysis, designing, coding, testing and documentation. Of course, the depth and detail of such a division is a secondary matter.

In the classical approach to the design process, steps are carried out sequentially. However, most software development methodologies are based on an evolutionary model as a consequence of static and incremental design [7]. In this approach, an important role is played by development laws determining the necessity and a form of the changes made in the design process and service as well and possible modifications of already-designed model. The evolutionary model is often presented in the form of a spiral [8], in which each of the mentioned stages corresponds to one of the cycle phases. The result of every proceeding stage is used as input for the next stage (Fig. 1) [9].

**Fig. 1** A classical software
development spiral

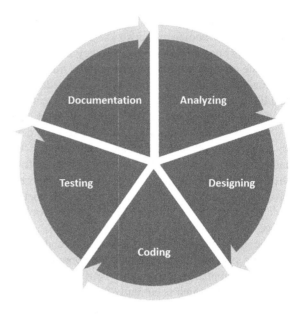

This approach also applies to the simulation and modeling process of computer networks and is related to the structuring of the design process, which depends on defining and scheduling tasks that should be made in order to obtain a specific solution. This is related to the requirements for future software. Thus, the structuring of the design process is expressed in the corresponding sequence of steps, sub-steps and individual tasks in the form of a system life cycle. Thus, the process of creating software for computer networks simulation and modeling may be also presented in the form of the following functional spiral (Fig. 2).

Of course, Fig. 2 presents the basic modules included in comprehensive software for modeling and simulation of computer networks. These modules include:

**Network resources allocation**. Functionality is one of the basic components of software for computer network modeling, it is associated with the allocation of individual components in order to achieve the most efficient operation of the infrastructure. Of course, the specific form of the allocation tasks is associated with the task of optimization, in which optimization criteria are selected and defined depending on needs e.g. cost, distance, capacity, reliability, etc. In the literature, multiple solutions of allocation tasks can be found, in particular: assignment task [10], multicriterial assignment task [11], quadrature assignments task [12], as well as hierarchical morphology design, based on the morphology clique [13], etc.

**Designing of physical topology**. Under the concept of network resources, the authors understand both the intermediary devices (e.g. switches, routers) as well as end devices (servers, databases, IoT elements). This task involves creating a set of physical connections that ensure the integrity of the network infrastructure, taking specific environmental conditions into account. Within this task, the available functionality should also provide the possibility to deliver redundant connections

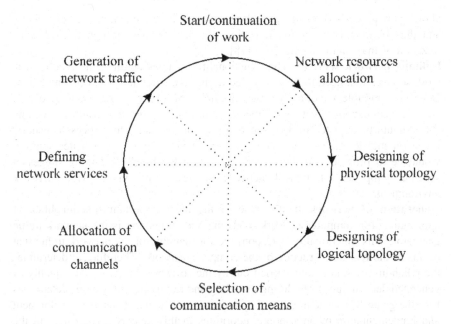

**Fig. 2** The spiral of software development process for modeling and simulation of computer networks

and consider reliability parameters. There is a wide range of algorithms supporting the designing of physical topology based on graph theory [14]. Of course, they are based on different models: capacity, incapacity, etc. [15].

**Designing of logical topology.** In this case, a connection structure to communicate between various network devices should be created. For this purpose, all currently known design methods can be used, including accurate and approximate methods (linear, nonlinear, dynamic, evolutionary, etc.) [16].

**Selection of communication means.** From the perspective of the distributed system performance—expressed by its speed, reliability, scalability or reconfiguration —the connection system is a significant part of the communication system, which is a set of specific communication means dedicated to specific tasks. Therefore, the proper choice of communication means should be one of the basic components involved in the designing of computer networks [17]. The task of selecting communication means may include: selection of transmission channels, selection of equipment or selection of channels and equipment. This selection may be implemented based on an existing base topology or with the designing of the connection structure topology.

**Allocation of communication channels.** The activities related to this functionality is to define the required (e.g. maximum, minimum, redundant) number of communication channels which results from the selected transmission media capabilities as well as throughput and reliability needs for the designed or simulated infrastructure. Selection of a wide range of available optimization algorithms can allow to obtain desired result e.g. in the form of limiting the number of transceivers and thus limiting costs or maximization of total channel throughput and minimization of maximum channel load [18].

**Defining network services**. This functionality includes defining the type of network services that need to be considered during the computer network simulation. Due to the possible use of a given network infrastructure, e.g.: data center network, convergence network, network supporting production process control, etc. this stage should be adjusted to their specificity of functioning. This process is usually based on the designer's knowledge and experience. However, a better solution would be to provide also the so-called "Knowledge base" containing suggested schema, templates of typical network services used in a given network environment.

**Generation of network traffic**. One of the most important functionalities of applications for computer network modeling and simulation is the network traffic generator. This functionality is responsible for mapping of the network traffic that occurs in the real infrastructure of the computer network. Therefore, it determines the reliability of simulations carried out by the software. Moreover, the quality of generated data streams used during modeling and testing, in many cases determines how the given ICT system will be implemented in practice. Thus, the development and implementation of appropriate algorithms in this field is a challenge. In the literature, there are many models and mechanisms for modeling of network traffic. The oldest model used was the Poisson distribution [19]. Another solution used for simulation of network traffic is Weibull or Pareto distribution [20]. One of the most famous is the Markov chain [21].

Of course, the scope and order of the individual functional areas depends on individual needs. It is associated with combination of certain modules in order to achieve a specific result, e.g. designing logical structure together with the selection of channels and end devices. In addition, the depth of the decomposition of the design process usually depends on the designer's knowledge and experience.

The presented approach, in line with most of currently applied software development methods, is based on the **sequencing and independence of each design phase**. It is characteristic for processes in simple systems. While the creation of software for modeling and simulation of computer networks should be based on the theory of complex systems [22].

## 3   Software Development Process in the Context of Complex Systems

Initially, complex systems theory referred primarily to biology. Over time, different scientific disciplines have begun to exploit achievements in this field for their purposes. These activities allowed them to interpenetrate, complement and modernize models and laws. The influence of social sciences such as sociology and economics is also important with this regard [23]. As a result, papers related to fractal geometry and chaos theory gradually appeared [4]. It can be concluded that these works were one of the pillars of Dynamic Systems Theory. On the other hand, papers related to self-organization [24], emergencies [25], autopoietal systems [26], small world theories [27], scale-free networks, power law [28], etc., allowed to understand the functioning of many natural and artificial systems [29]. Thereupon, the complex systems theory continuously finds new application and also describes the mechanisms and phenomena that exist in many different areas.

One of the basic properties of complex systems is the phenomenon of feedback loop. Feedback is directly related to the self-adaptation of complex systems. In the case of software development for simulation and modeling of computer networks, feedback loop allows to customize a created software to changing conditions, requirements and information about possible errors. It should be noted that the feedback loops in complex systems may have nested character of different scale effect on the system [30]. Thus, micro- and macro-scale dependencies are reflected in them. These relations have been used to propose a new approach to the software development process for modeling and simulation of computer networks (Fig. 3).

This approach assumes the need of taking into account all designed and developed functionalities throughout the development cycle. This principle is especially important when creating software for modeling and simulating computer networks, which components may contain algorithms combining functionalities defined in separate modules to smaller or greater extend. Individual circles (marked

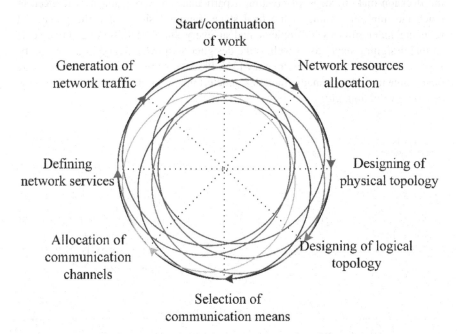

**Fig. 3** New spiral of software development process for modeling and simulation of computer networks

in different colors) represent the essence of the feedback loop for specific functional modules. Adoption of such a concept forces changes in existing ones or incorporating this into newly developed software methodologies. As a result, while designing every functional module, retrospection should be made in the context of the existing and planned modules. In case of justified necessity, already created modules should be inspected and changes need to be introduced in them. This approach makes it impossible to assume in advance that the functional module has a final terminated statute. Thus, this property is in contradiction to the current approach found in Agile methodologies, where the created and tested module should be ready for final delivery to the client. This approach is presented in Fig. 2, where individual functional modules are implemented independently. They can be created sequentially or in parallel, however, without being coupled. Moreover, in the proposed solution, it is possible to improve the functional module as a result of its analysis from the point of view of microscopic feedback. However, the macroscopic approach enables a correlation between all the other modules throughout the life cycle of the software being developed.

In addition, indeterminacy is significantly reduced during designing and implementation process that allows to develop more effective solutions. This issue will be discussed later in this paper.

While maintaining an adequate level of structural design process, it can be assumed that this design is actually the process of reducing the indeterminacy of the object being designed. It should be noted that the greatest indeterminacy occurs at the decision-making stage of creating a particular software, then it is reduced to reach the minimum value at the final stage. Figure 4 shows that the process of reducing indeterminacy is implemented at every stage of designing. Of course, formal designing methods as well as the designer's knowledge and intuition can be used in the reduction of indeterminacy [31]. Please note that in the case of software development, indeterminacy includes indeterminacy of input data, technology, criteria, evaluation, etc.

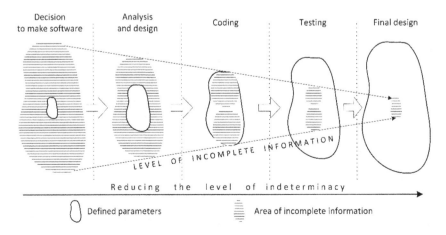

**Fig. 4** Reduction of indeterminacy in the software development process

Existence of indeterminacy is closely connected with complex systems, where every attempt of the description formalization in the form of models is connected with more or less generalization. Thus, the process of pursuit to refine the model is a limitation of indeterminacy. Another feature of complex systems e.g. feedback loop plays an important role.

One of the important properties of complex systems is the non-additivity of their processes [32]. Of course, this paper does not attempt to focus on the non-additive processes occurring in the environment where a given software is running and the dependencies between the hardware, algorithm and the input data associated with them. Non-additivity includes two areas of software development:

**Functional**—in this case the sum of the individual functional areas is something more than just a simple connection. Therefore, designing and modeling of network functioning by means of separate functional modules will not allow to obtain effective solution. This is due to the fact that in many cases there is a close link between the particular design stages, such as the creation of physical and logical topology and the selection of appropriate communication means, etc. Thus, consideration of this type of additivity is required at the stage of software designing and development.

**Process**—is the second case of non-additivity during software development (Fig. 5), where stages such as designing, coding, and testing should be implemented simultaneously in many cases. Then, the value resulting from this operation is greater than the common sum of the effects of the particular stages of software design and development.

Including indicated non-additivity in the software development process enables:

- faster detection of software bugs,
- better code optimization in respect of e.g. tested hardware and software platforms,
- better customization,
- costs reduction associated with later modifications of software functional structure, etc.

Especially, in the case of software for modeling and simulation of computer networks, maintaining non-additivity seems justified. Such software is based on mathematical models to a great extent that describe the behavior of the computer network itself, the phenomena taking place in it, but also the mechanisms for optimal design of network infrastructure. In this case, the correlation between the particular stages of software development or the teams implementing them

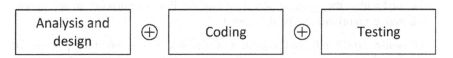

**Fig. 5** Non-additivity process

significantly influences the effectiveness of the end solution. Keep in mind that even a small modification to speed up the algorithms can result in a significant reduction of the time needed to compute the final solution.

Software used to model and simulate computer networks should have a wide range of functionalities. Only then, it is possible to provide the necessary tools for designers, architects, scientists and computer network administrators. An example of basic functional groups is shown in Figs. 2 and 3. Among them are modules: network resources allocation, design of physical and logical topology, selection of communication means, etc. Of course, due to the complexity of such software, solutions are created that support a limited functional range, often only one group. This paper presents the possibility of using mechanisms, features and properties associated with complex systems in the process of creating software for particular functional groups. Taking them into account can help to create software for modeling and simulation of computer networks, which will considerably take into account the processes and phenomena taking place in the real network infrastructure. Below are presented a few selected features/properties of complex systems that should be included in the process of creating the indicated functional modules:

- **Self-similarity** was officially recognized in network traffic in 1994 [33], as a fractal concept in data analysis, and some sources [34] reported it as early as in 1990. It allows to create algorithms for simulation of network traffic allowing scaling its course in different time bands. This functionality allows to make the algorithm of generating network traffic independent from the time of the simulation set by the user.

- **Power law and scale-free network** indicate that there is no proportional distribution in reality, but some elements have much more connections and relationships than others [28]. On the basis of this law, an algorithm was created which allows to model the scale-free networks. This algorithm and its various modifications can be used in the modules of creating physical and logical topology. They can be perfectly used for modeling wireless networks and hierarchical structures in wired networks. They also allow to make processes involved in making connections between network elements more realistic, both in the capacity and incapacity models [15].

- **Small-world networks** allow to get to any destination in a relatively small number of traffic hops [27]. This property can be used in most software modules connected with modeling and simulation of computer networks. Obviously, the primary area of application are modules for creating the physical and logical connections, but also generating network traffic on the basis of applications simulating relationships in social groups. This feature of complex systems can be used within the resource allocation module. For this purpose, an appropriate clustering algorithm can be developed.

Of course, above indicated properties of complex systems and their potential of application during creating specific software modules for modeling and simulation of computer networks are only a small example. Obviously, the remaining features of complex systems such as emergence, nonlinearity, etc. can also be used for this purpose. However, this issue should be the subject of a separate more complex publication.

# 4 Examples of Application

Software for computer network modeling and simulation can have many independent and interoperable functional modules. Some selected elements are presented below, which confirm the validity of above described approach of software development.

The first element is the module for generating network traffic. In course of tests, it turned out that the network traffic patterns used so far and based on distributions such as Poisson's have not corresponded to the real situations encountered in computer networks [35]. Therefore, tests for different network traffic models were performed. It turned out that the best results were achieved by the scalability of registered network traffic in real network infrastructure based on the fractal model described by means of non-extensive statistics. Numerous studies on self-similarity of real network traffic based on Hurst coefficient and other statistical parameters have been conducted. In this case, the studies included both classic TCP and UDP flows, but also the traffic generated by different network applications was checked. All these works confirmed the self-similarity of real network traffic, and Hurst coefficient was in the range of 0.5–1 [36]. Therefore, the use of fractal algorithms in the network traffic generation module is fully justified.

In the second case, the possibility of creating software modules based on the concept presented in Fig. 3 in the form of multiple spirals was investigated. This approach was used for modeling broadband network for Podkarpacie. For this purpose, modules were developed for the allocation of network resources, the creation of physical and logical topologies as well as the selection of communication means. Figure 6 shows the implementation of the multi-spiral model discussed in this case. At every key stage, a local feedback loop was made in the design process to adjust the software in order to meet changing needs. Global feedback was applied in two cases: the correlation with topological logic with physical topology design as well as the allocation of communication channels and the design of logical topology. The adopted approach allowed for a technically coherent solution.

Due to the complexity of software for computer network modeling and simulation, the development of complete software requires a large effort and a team of specialists. Therefore, only a certain range of implemented functionality is presented. However, the results, conclusions and observations confirm validity of the approach presented by the authors.

**Fig. 6** Example of new
spiral application

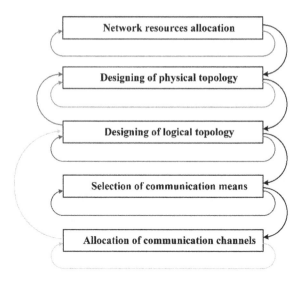

## 5    Conclusions and Further Work

Designing and creating a software used to model and simulate the operation of
computer networks is a major challenge for system architects, designers and
developers. Due to the complexity of the phenomena that need to be programmed
(among others real conditions of computer network operation), models and
mechanisms are being constantly searched. Taking into account the large correlation of complex systems models, it seems to be reasonable to use this theory for this
purpose.

The authors presented a look at the process of creating software for modeling
and simulation of computer networks in terms of complex systems. The concept of
feedback loops was used to modify the spiral model. In addition, the process of
minimizing indeterminacy in the software development process has been shown.
This is a natural process in case of creating advanced and extensive software. An
important aspect was also to indicate non-additive stages of designing and developing process. This concept can contribute to improvement of the performance of
software development process and better correlation between the individual
modules.

It should be clearly stated that the presented paper is an introduction to the use of
complex systems theory while creating software. The main purpose of this paper
was to outline that the software development process (especially for computer
network modeling and simulation) should be treated the same as the processes in
the complex systems. Due to the extent of the subject matter, all the important
aspects cannot be presented in one paper. Therefore, according to authors, the
following issues need to be addressed in the future:

- Development of a full software development methodology based on selected properties of complex systems, including functional and process non-additivity and feedback loops.
- Development of tools to support project management based on the new methodology.
- Development of methods and measures to reduce indeterminacy in the software development process.
- Study of the emergence properties in the processes accompanying software development.
- etc.

We hope this paper will contribute to an in-depth discussion of the role of complex systems theory in software engineering.

# References

1. Anthonysamy, P., Rashid, A.: Software engineering for privacy in-the-large. In: Proceedings of the 37th International Conference on Software Engineering, vol. 2, pp. 947–948. IEEE Press Piscataway, NJ (2015)
2. Sethi, A.S., Hnatyshin, V.Y.: The Practical OPNET User Guide for Computer Network Simulation. CRC Press (2013)
3. Varga, A., Hornig, R.: An overview of the OMNeT++ simulation environment. In: Proceedings of the 1st International Conference on Simulation Tools and Techniques for Communications, Networks and Systems and Workshops, ICST, Brusskes (2008)
4. Nicolis, G., Nicolis, C.: Foundations of Complex Systems: Emergence, Information and Predicition. World Scientific Publishing Co., Singapure (2012)
5. Shooman, M.L.: Reliability of computer systems and networks. In: Fault Tolerance, Analysis and Design. Wiley, New York (2002). doi:10.1002/047122460X
6. Cormen, T.H., Leiserson, ChE, Rivest, R.L., Stein, C.: Introduction to Algorithms. MIT Press and McGraw Hill, New York (1990)
7. Fisher, V.G.: Evolutionary Design of Corporate Networks Under Uncertainty. Technischen Universität München, München (2000)
8. Hajder, M., Paszkiewicz, A., Bolanowski, M.: Intuicja projektanta jako niejednoznaczność przy projektowaniu sieci komputerowych. Współczesne problemy sieci komputerowych – nowe technologie, Wydawnictwa Naukowo-Techniczne, Gliwice, pp. 259–268 (2004)
9. Boehm, B.W.: A Spiral model of software development and enhancement. Computer 21(5), 61–72 (1988)
10. Garey, M.R., Johnson, D.S.: Computers and Intractability. The Guide to the Theory of NP-Completeness. W.H. Freeman & Company, San Francisco (1979)
11. Current, J., Min, H., Schilling, D.: Multiobjective analysis of facility location decisions. Eur. J. Oper. Res. 49(3), 295–307 (1990)
12. Cela, E.: The Quadratic Assignment Problem: Theory and Algorithms. Kluwer Academic Publishers, Dordrecht (1998)
13. Levin, M.: Composite Systems Decisions. Springer, New York (2006)
14. Derrible, S., Kennedy, Ch.: Applications of graph theory and network science to transit network design. J. Transp. Rev. 31(4), 495–519 (2011)
15. Claus, A., Kratzig, S.: Optimal planning of network structures within an exchange area. Eur. J. Oper. Res. 7(1), 67–76 (1981)

16. Puech, N., Kuri, J., Gagnaire, M.: Models for the Logical Topology Design Problem. Springer, London (2002)
17. Hajder, M., Paszkiewicz, A.: Selecting communication means in condition of incomplete information for distributed systems with hierarchy of topology. Polish J. Environ. Stud. **17**(2A), 19–23 (2008)
18. Bolanowski, M., Paszkiewicz, A.: Reconfiguration of logical communication channels. J. Theoret. Appl. Comput. Sci. **7**(3), 21–28 (2013)
19. Paxson, V., Floyd, S.: Wide-area traffic: the failure of poisson modeling. IEEE/ACM Trans. Netw. **3**(3), 226–244 (1995)
20. Chandrasekaran, B.: Survey of Network Traffic Models. Washington University in St, Louis, St. Louis (2015)
21. Gebali, F.: Analysis of Computer Networks. Springer International Publishing (2015). doi:10. 1007/978-3-319-15657-6_16
22. Braha, D., Minai, A., Bar-Yam, Y.: Complex Engineered Systems. Science Meets Technology. Springer, New York (2006)
23. Boulding, K.E.: General systems theory: the skeleton of science. Manage. Sci. **2**(3), 197–208 (1956)
24. Bertelle, C., Duchamp, G.H.E., Kadri-Dahmani, H.: Complex Systems and Self-Organization Modelling. Springer, Berlin, Heidelberg (2009)
25. Kron, T., Grund, T.: Society as a selforganized critical system. Cybern. Hum. Knowing **16**(1–2), 65–82 (2009)
26. Hall, W.P., Nousala, S.: Autopoiesis and knowledge in self-sustaining organizational systems. In: 4th International Multi-Conference on Society, Cybernetics and Informatics: IMSCI 2010, Orlando, Florida, 29 June–2 July 2010 (2010)
27. Easley, D., Kleinberg, J.: Networks, Crowds, and Markets: Reasoning About a Highly Connected World. Cambridge University Press (2010)
28. Barabási, A.L.: Scale-free networks: a decade and beyond. Sci. Mag. **325**, 412–413 (2009)
29. Grabowski, F., Paszkiewicz, A., Bolanowski, M.: Wireless networks environment and complex networks. In: Analysis and Simulation of Electrical and Computer Systems, Lecture Notes in Electrical Engineering, vol. 324, pp. 261–270. Springer International Publishing (2015)
30. Aström, K.J., Albertos, P., Blanke, M., Isidori, A., Schaufelberger, W., Sanz, R. (eds.): Control of Complex Systems. Springer, London (2001). doi:10.1007/978-1-4471-0349-3
31. Hajder, M., Paszkiewicz, A.: Indeterminacy in the Systems and Networks Design, pp. 269–278. Annales UMCS Sectio Ai Informatica, Wydawnictwo UMCS (2004)
32. Grabowski, F.: Nonextensive model of self-organizing systems. Complexity **18**(5), 28–36 (2013)
33. Leland, W.E., Taqqu, M.S., Willinger, W., Wilson, D.V.: On the self-similar nature. IEEE/ACM Trans. Netw. **2**(1), 1–15 (1994)
34. Wilson, M.: A Historical View of Network Traffic Models. https://www.cse.wustl.edu/~jain/cse567-06/ftp/traffic_models2/ (2017). Accessed 24 May 2017
35. Bolanowski, M., Paszkiewicz, A.: The use of statistical signatures to detect anomalies in computer network. In: Analysis and Simulation of Electrical and Computer Systems, Lecture Notes in Electrical Engineering, vol. 324, pp. 251–260. Springer International Publishing (2015)
36. Domańska, J., Domańska, A., Czachórski, T.: A few investigations of long-range dependence in network traffic. In: Proceedings of the 29th International Symposium on Computer and Information Sciences, pp. 137–144. Springer International Publishing (2014)

# A Scrum-Centric Framework for Organizing Software Engineering Academic Courses

Mirosław Ochodek

**Abstract** Teaching Scrum and other complex Software Engineering (SE) practices, methods, and tools within a regular academic course is often a challenging task because the examples shown to students and their working environment are not realistic enough. This problem is frequently tackled by organizing courses in the form of capstone projects. Unfortunately, this approach requires additional resources (e.g., more tutors, external customers, etc.) what limits the potential number of participants. As a response to this problem, we propose a Scrum-centric framework that allows combining lectures and laboratory classes with a minimalistic capstone project, all within a regular SE course. The course is organized similarly to a project run according to the Scrum guidelines. The focal point of the framework is the synchronization between the content presented during lectures and laboratory classes with the capstone project iterations (sprints). We also share our experience from 2 years of conducting Software Engineering course organized according to the framework together with the results of a cross-sectional survey assessing student perceptions of the effectiveness of the approach.

## 1 Introduction

Software Engineering (SE) plays an important role in Computer Science education. The SE courses are present in ACM/IEEE curricula for Computer Science [2], Computer Engineering [3], and Information Technology [1].

Among the many SE areas covered by IEEE/ACM curriculum for Computer Science, there are some related to software development processes and project management. Taking into account current trends in software development (SD), it seems valuable to teach students how to apply Agile SD methods, and in particular, Scrum [15], which appears to be the most popular Agile SD method nowadays [5].

M. Ochodek (✉)
Poznan University of Technology, Institute of Computing Science,
ul. Piotrowo 2, 60-965 Poznan, Poland
e-mail: mochodek@cs.put.poznan.pl

© Springer International Publishing AG 2018
P. Kosiuczenko and L. Madeyski (eds.), *Towards a Synergistic Combination of Research and Practice in Software Engineering*, Studies in Computational Intelligence 733, DOI 10.1007/978-3-319-65208-5_15

Unfortunately, teaching students how to apply Scrum within a regular Software Engineering course consisting only of lectures and laboratory classes is a challenging task. Of course, it is possible to present and discuss Scrum roles, events, and artifacts and even practice them during laboratory classes. However, it seems difficult to provide students with deeper understanding of the concepts like applying empirical process control (transparency, inspection, and adaptation), delivering business value to customers, or showing the role of technical debt in software maintenance. We believe that similar problems could be also observed in the case of teaching other complex SE practices, methods, and tools.

The aftermentioned problem is not new, and there at least 2 approaches to tackle with it. The first one is to use different mini-games or workshops that use metaphors of software projects to help students understand Scrum or other practices (e.g., [14, 16, 17]). Although such games are useful and attractive, the project environments they create are still artificial. Therefore, a more commonly used approach is to organize capstone projects or software development studios [6, 7, 10, 11]. In these frameworks, teams of students deliver software products for external customers. They create excellent opportunities to practice application of different SE practices, methods, and tools (including SD methodologies like Scrum). Unfortunately, these approaches have some drawbacks as well. Firstly, capstone projects require additional resources to simulate a "real" project environment [10]. Therefore, they are difficult to organize within a regular SE course. Secondly, students need to have a basic knowledge regarding SE to effectively participate in such a course. Consequently, capstone projects are usually organized as specialized, follow-up courses. Finally, our experience shows that such courses do not scale well with respect to the number of students. A typical Scrum capstone course presented in the literature has on average 25 participants [6, 7, 11] while, for instance, at Poznan University of Technology we have more than one hundred B.Sc. students attending Software Engineering course each year.

Therefore, the question arises: *how to incorporate a Scrum capstone project into a regular Software Engineering course consisting only of lectures and laboratory classes?*

We address this question by proposing a Scrum-centric framework for organizing a Software Engineering B.Sc.-level course that aims at synergizing benefits of regular lectures, laboratory classes, and a minimalistic capstone project. In particular:

- we discuss the current approaches to teaching Scrum, and in particular organizing capstone courses (see Sect. 2);
- we present a framework that transforms a regular Software Engineering course into a project run according to Scrum guidelines (see Sect. 3);
- we share the lessons learned from applying the framework within the B.Sc.-level Software Engineering courses at Poznan University of Technology and present the results of an end-of-term student opinion survey concerning the effectiveness of the course (see Sect. 4).

## 2 Related Work

The problem of how to teach Software Engineering, and in particular, Scrum has been considered by numerous studies.

The most compact way of teaching Scrum is to use one of the existing Scrum games, such as DELIVER! [16], Scrumia [17], or Virtual Scrum [14]. These methods illustrate the iterative process used in Scrum in an attractive way. They also introduce the idea of retrospective and continuous improvement. The methods are valuable tools for tutors and attractive form of learning for students. However, they are still not exactly accurate in showing the true nature of SD projects.

A more realistic approach to teaching Scrum and other Software Engineering practices, tools, and methods is to organize courses in the form of capstone projects or software development studios [6, 7, 10, 11]. In such courses, groups of students are developing software products for external customers. In the simplest variant, a tutor is trying to roleplay the real customer, however, in many cases, projects are developed for external customers (companies, organizations, etc.), often willing to pay for the products [10].

Capstone courses are conducted as a follow-up to regular courses on Software Engineering [6, 13] to assure that students have already acquired knowledge and skills necessary to participate in a software development project. In addition, they are often organized as multi-semester courses [6, 10, 13]. They also require additional resources and infrastructure, what limits the number of students attending the classes. We found that the number of participants of such courses described in the literature ranged between 5 and 52 (mean = 25) per semester [6, 7, 11]. The main difference between these studies and the framework presented in this paper is that we are trying to incorporate a minimalistic capstone project within a regular course that could be attended by more than hundred students.

Finally, capstone courses can also include some instructional classes [6, 7, 9, 11, 12]. However, their role is auxiliary and the number of hours is limited. In our approach, lectures and laboratory classes play the key role, while capstone projects have a supportive role of illustrating and practicing the content presented during the lectures and laboratory classes.

## 3 Scrum-Centric Course Framework

The proposed framework assumes three forms of classes: regular lectures, laboratory classes, and participation in a capstone project. The focal point of the framework is the synchronization between all those forms of teaching. The main idea is that during the sprint $N$, students learn some Software Engineering practices, methods and tools by participating in lectures and laboratory classes. The content they have learned during the sprint $N$ is then added to the Definition of Done (DoD) as technical requirements for the next sprint $N + 1$.

## 3.1  Learning Objectives

According to Bloom's revised taxonomy [4] the learning objectives can be divided into five categories: (1) remember, (2) understand, (3) apply, (4) evaluate, and (5) create. Together, they form a logical sequence of the learning process. For instance, to understand something you have to remember it. Similarly, you will not be able to apply something until you understand it. At the bachelor level curricula, the main focus is usually on the learning objectives belonging to the first three categories of the taxonomy (remember, understand, and apply).

The framework is designed to support two types of learning objectives. The first group concerns Scrum, while the second group regards specific SE practices, methods, and tools being in the scope of the course. Here, we only define learning objectives for the Scrum, assuming that the learning objectives for the specific content of the course depend on its program.

The proposed learning objectives for Scrum are defined at two levels—basic and advanced. The basic level objectives are as follows:

- Scrum roles—remember, understand, and practice performing Scrum roles.
- Scrum events—remember, understand, and practice organizing and participating in Scrum events (Sprint Planning, Daily Scrum, Sprint Review, and Sprint Retrospective).
- Scrum artifacts—remember, understand, and practice creating and maintaining artifacts such as Product Backlog, Sprint Backlog, etc.
- Scrum rules—remember, understand, and practice Scrum rules and process.

  The advanced learning objectives regard the following aspects of Scrum:

- Acceptance criteria, Definition of Done (DoD), and potentially shippable increment of a product—remember, understand, and practice defining and delivering software that meets certain acceptance criteria and technical requirements imposed by DoD.
- Empirical process control—remember, understand, and apply the empirical process control proposed by Scrum (transparency, inspection, and adaptation).

## 3.2  Course Organization

We propose to divide the course into a series of sprints (iterations) as it is presented in Fig. 1. The figure shows the main activities as well as the synchronization between lectures, laboratory classes, and project sprints.

The process starts with a *Scrum introductory lecture*. Its goal is to familiarize students with the values, roles, artifacts, events, and rules of Scrum. During the first laboratory classes, students form teams of four (in exceptional cases, teams of three) and *formulate problems* they would like to solve by building software products. Finally, they assign the role of Scrum Master and Product Owner for the first sprint of their projects.

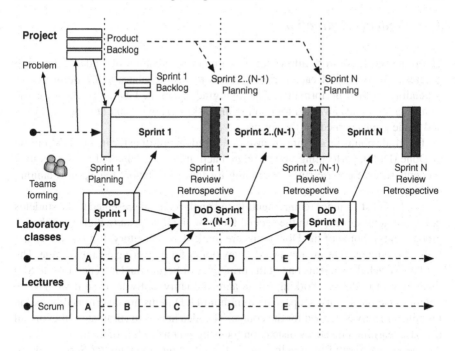

**Fig. 1** Course process showing synchronization between lectures, laboratory classes, and project

Meanwhile, the teams start to prepare their initial *Product Backlogs*. The elected Product Owner in each project is in charge of that process. The Product Backlog needs to be established until the classes dedicated to the planning of the first sprint.

During the first *Sprint Planning* classes students are presented with Definition of Done for the first sprint and other requirements imposed on the teams regarding process and product. The outcome of the classes is the refined Product Backlog and *Sprint Backlog* for the sprint. The sprint starts.

We assume that the duration of sprint is 4 weeks. During the sprint, we dedicate one laboratory classes for *teamwork*. It allows students to work together in one place and also practice *Daily Scrum* meetings or other team activities, such as Pair Programming. Nevertheless, we perceive these classes as optional. They could be introduced depending on how much content is planned for the course.

The sprint ends with laboratory classes regarding *Sprint Review* and *Sprint Retrospective*. The teams present, in front of the class, the increments of products they developed during the sprint. They also discuss the further directions of the development of their products. Finally, they perform a retrospective meetings to determine improvement actions for the next sprint. If possible, teams also prepare a Sprint Backlog for the next sprint.

This cycle is repeated until the end of the semester. Within a course lasting for 15 weeks, it should be possible to perform at least three sprints. The "cost" of implementing the framework is equal to at least $N + 2$ units of laboratory classes (Scrum events, a lecture on Scrum, and team forming) up to $2 \times N + 2$ if one would like to introduce additional classes for working in teams.

## 3.3  Students Evaluation

The proposed framework allows for a continuous evaluation of students' learning progress. At the end of each sprint, students receive feedback and *grading points* depending on their performance. The final grade from the laboratory classes is the sum of grade points awarded for completing tasks during regular laboratory classes and grade points earned for the sprints.

The grade points for a sprint are awarded based on fulfilling the items of Definition of Done (DoD), performing Scrum-related activities, and delivered business value. Students receive precise rules of grading for the sprint before the Sprint Planning meeting.

The DoD list includes requirements based on the content presented to the students in the previous sprints, with the special focus on the last sprint (a higher number of grade points). Some of the DoD items are marked as obligatory. If a team does not fulfill those items, they are not awarded any points for delivering business value. The goal is to teach the students one of the principles of Scrum (and in general Agile SD) that we want to deliver working (of an agreed quality) software even if we have to compromise the sprint's scope. The minimum DoD items are determined based on learning objectives defined for the course. The idea here is to stimulate achieving all the base learning objectives instead on focusing on one or few of them.

The second part of the grading list is the set of requirements for Scrum-related activities. Students are expected to organize Scrum events (Scrum Planning, Retrospective, etc.), regularly update their Sprint Backlog, implement actions from Retrospectives, etc. They earn grading points for correctly performing each of those activities.

The last category of the grading list relates to delivered business value. The number of grading points depends on the number of "valuable" elementary functions provided in the increment of the software product. The term "valuable" means that the functions allow users to obtain some goals (maybe, for the time being, in a very simplified way). A good example of a mistake that students often make is to start implementation from functions like login or register user, which in the majority of cases could be easily postponed until the later increments.

Finally, all team members anonymously vote on how to divide the earned grade points depending on the team members' contribution to the sprint. If needed, the decision can be refined by the tutor.

## 4  Validation of the Framework

A course designed according to the proposed framework has been conducted at Poznan University of Technology (PUT) for nearly 3 years now. The name of the course is Software Engineering II. It is a second semester of the two-semester course on

Software Engineering at B.Sc.-level. Each year it is attended by 100–130 students in their 6th semester.

The course consists of $15 \times 90$ min of lectures and $15 \times 90$ min of laboratory classes. The expected workload of each student is 125 h (5 ECTS points). The course covers software development methodologies and project management (Scrum, XP, PRINCE2, project planning, etc.), software testing (unit, integration, acceptance, and non-functional testing) and software quality (software metrics, GQM, code refactoring), configuration management (automatic builds, Continuous Integration), and formal methods.

The software infrastructure provided to students includes Redmine used for tasks and backlogs management, Subversion, and Jenkins to support Continuous Integration. We also developed an application allowing students to track their progress and grade points they earn.[1]

## 4.1 Survey Research

As an element of continuous improvement of the course, at the end of each semester, we conduct a cross-sectional survey assessing student perception of the effectiveness of the course organizing according to the proposed framework for achieving learning objectives in comparison to regular classes (RQ1). In addition, we investigate how the core mechanisms of the framework affect the motivation of the students to learn (RQ2).

### 4.1.1 Survey Design

The designed survey consisted of two main parts, each related to one of the research questions. We also added an open question at the end of the survey to allow students expressing their general remarks and observations.

The first part includes 21 statements related to learning objectives presented in Sect. 3.1. The first 18 statements regard learning objects for Scrum (remembering, understanding, and applying). The remaining three statements refer to learning objectives for the regular SE content of the course. The general template of these statements could be defined as follows:

*Participation in a capstone project in addition to lectures and laboratory classes allowed me to better <remember | understand | apply> <the name of a learning objective> than if the classes would consist of lecture and laboratories only.*

---

[1]http://io.cs.put.poznan.pl.

The students are asked to express they (dis)agreement with the statements using a five-point Likert scale (definitely agree, rather agree, hard to say, rather disagree, definitely disagree).

The second part of the survey aims at investigating how different mechanisms of the framework affects the motivation of the students. For the purpose of characterizing "motivation", we use the ARCS model [8]. The model defines four conceptual categories characterizing human motivation:

- *attention*—the learning material should gain person's attention and sustain it throughout a period of instruction;
- *relevance*—a person has to feel that the presented material is useful;
- *confidence*—a person should feel that the success is possible if effort is exerted;
- *satisfaction*—a person should feel satisfied with his/her achievement (e.g., feel rewarded appropriately).

We define two statements per each category (eight statements in total). All of them are answered using the same five-point scale. The first type of the statement had a form:

> *The form of conducting classes <was more attractive | kept me more engaged | gave me more confidence in being able to achieve my goals | kept me more satisfied> than regular classes consisting of lectures and laboratory classes only.*

The second type of statements listed different mechanisms of the framework and their influence on different categories of the ARCS model.

> For each of <attention | relevance | confidence | satisfaction>:
>
> - *the possibility of trying out in practice the SE practices, tools, and methods presented during lectures and laboratories;*
> - *the possibility of applying Scrum in a project;*
> - *working in a team;*
> - *being challenged by DoD items.*

To reduce the number of items, we do not ask about the "confidence" category for items related to applying SE or Scrum in practice because we perceive it as irrelevant in their context.

## 4.2 Survey Results and Discussion

The survey was administered online to 2 years of students at the end of each semester (2015 and 2016). We received 146 anonymous responses (68 in 2015 and 78 in 2016). The response rate was greater than 60%.

The distribution of responses regarding learning objectives is presented in Fig. 2. For further investigation, we simplified the scale by transforming it into a two-point

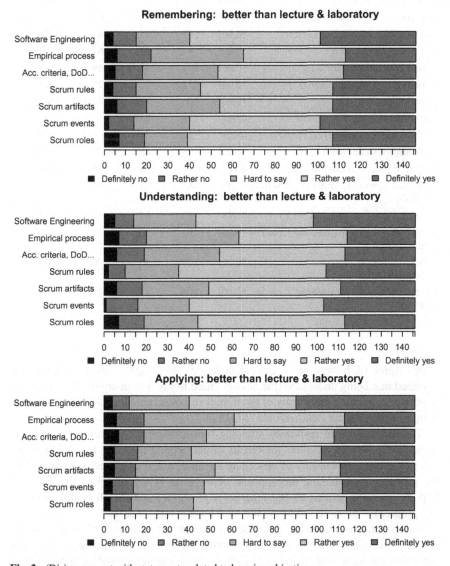

**Fig. 2** (Dis)agreement with statements related to learning objectives

**Table 1**  Summary of the responses for the statements regarding effectiveness in achieving learning objectives of the proposed framework

| Statement | p-value | agree | ¬agree | % agree | % ¬agree |
|---|---|---|---|---|---|
| Scrum roles (REMEMBER) | *0.001 | 107 | 39 | 73.3 | 26.7 |
| Scrum roles (UNDERSTAND) | *0.001 | 102 | 44 | 69.9 | 30.1 |
| Scrum roles (APPLY) | *0.001 | 104 | 42 | 71.2 | 28.8 |
| Scrum events (REMEMBER) | *0.001 | 106 | 40 | 72.6 | 27.4 |
| Scrum events (UNDERSTAND) | *0.001 | 106 | 40 | 72.6 | 27.4 |
| Scrum events (APPLY) | *0.001 | 99 | 47 | 67.8 | 32.2 |
| Scrum artifacts (REMEMBER) | 0.001 | 92 | 54 | 63.0 | 37.0 |
| Scrum artifacts (UNDERSTAND) | *0.001 | 97 | 49 | 66.4 | 33.6 |
| Scrum artifacts (APPLY) | *0.001 | 94 | 52 | 64.4 | 35.6 |
| Scrum rules (REMEMBER) | *0.001 | 101 | 45 | 69.2 | 30.8 |
| Scrum rules (UNDERSTAND) | *0.001 | 111 | 35 | 76.0 | 24.0 |
| Scrum rules (APPLY) | *0.001 | 105 | 41 | 71.9 | 28.1 |
| Definition of Done (REMEMBER) | 0.001 | 93 | 53 | 63.7 | 36.3 |
| Definition of Done (UNDERSTAND) | 0.001 | 92 | 54 | 63.0 | 37.0 |
| Definition of Done (APPLY) | *0.001 | 98 | 48 | 67.1 | 32.9 |
| Empirical process control (REMEMBER) | 0.107 | 81 | 65 | 55.5 | 44.5 |
| Empirical process control (UNDERSTAND) | 0.058 | 83 | 63 | 56.8 | 43.2 |
| Empirical process control (APPLY) | 0.028 | 85 | 61 | 58.2 | 41.8 |
| Other Software Engineering content (REMEMBER) | *0.001 | 106 | 40 | 72.6 | 27.4 |
| Other Software Engineering content (UNDERSTAND) | *0.001 | 103 | 43 | 70.5 | 29.5 |
| Other Software Engineering content (APPLY) | *0.001 | 106 | 40 | 72.6 | 27.4 |

∗ p-value smaller than 0.001

scale—agree (definitely agree and rather agree), and ¬agree (all the others). We assumed that being undecided (the answer "hard to say") is an argument against the hypothesis about the superiority of the framework over regular courses consisting of lectures and laboratory classes only (RQ1).

The distribution of responses after the transformation is presented in Table 1. In addition, the table presents the results of performing the exact binomial tests to determine if the probability of observing a response agree $p(\text{agree})$ is greater than the probability of making the opposite observation $p(\neg\text{agree})$. Formally, the null hypotheses were defined as $H_0$: $p(\text{agree}) = p(\neg\text{agree})$ and the alternative ones as $H_1$: $p(\text{agree}) > p(\neg\text{agree})$. The significance level $\alpha$ was set to 0.05.

Based on the results we can conclude that majority of students confirmed the superiority of the course organized according to the framework over traditional classes in achieving the basic-level learning objectives regarding Scrum and other SE-related

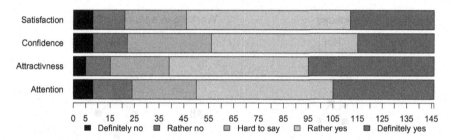

**Fig. 3** (Dis)agreement with the statements related to motivation

**Table 2** Summary of the responses for the statements regarding impact of the framework mechanisms on students' motivation to learn

| Statement | p-value | agree | ¬agree | % agree | % ¬agree |
|-----------|---------|-------|--------|---------|----------|
| Attention | *0.001 | 96 | 50 | 65.8 | 34.2 |
| Relevance | *0.001 | 107 | 39 | 73.3 | 26.7 |
| Confidence | 0.003 | 90 | 56 | 61.6 | 38.4 |
| Satisfaction | *0.001 | 100 | 46 | 68.5 | 31.5 |

∗ p-value smaller than 0.001

content (on average ca. 70% respondents agreed; all the null hypotheses rejected). For the learning objective regarding remembering, understanding, and applying empirical process control proposed by Scrum (advanced level), the results were less convincing (55.5–58.2% of "agree" responses; p-values greater than the assumed $\alpha$ for remembering and understanding). However, for the remaining advanced-level learning objectives, the average agreement was ca. 64%, and all the null hypotheses were rejected.

The summary of the responses regarding motivation (RQ2) is presented in Fig. 3. Again, we transformed the responses into the two-point scale agree/¬agree to verify if the mechanisms of the framework, in general, had a positive impact on the different aspects of motivation. The results of this step are presented in Table 2. All the observed differences were statistically significant (the average number of responses "agree" was ca. 67%).

In the following step, we analyzed statements regarding the impact of specific mechanisms on students' motivation, and transformed the responses into a three-point scale (yes, no, and hard to say). We decided to distinguish between "no" and "hard to say" this time because the former means that the students found this mechanism disturbing, while the latter response indicates neutral attitude. The summary of the responses, in the form of radar charts, are presented in Fig. 4.

The strongest impact on different aspects of motivation was observed for "applying the SE practices, tools, methods in practice". Similar results were observed for "practicing Scrum", however, in this case, we observed a visible group of students who did not perceive it as a factor increasing their attention. The other factor that visibly affected motivation was the "possibility of working in a team". It had the

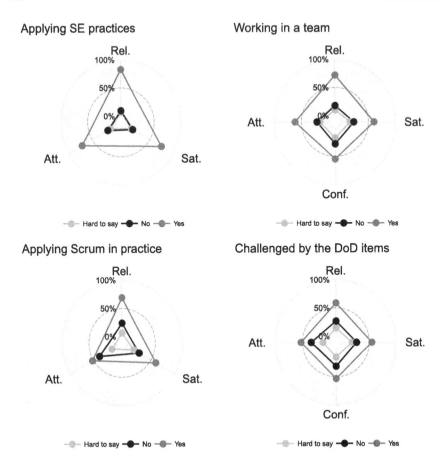

**Fig. 4** Influence of the framework mechanisms on the ARCS model categories

biggest impact on making the course relevant to the students. Working in a team also had a positive impact on keeping engaged (attention) and the feeling of satisfaction. The idea of introducing "Definition of Done that challenges students" also seem to impact their motivation (especially in the case of satisfaction and relevance).

## 4.3  Threats to Validity

There are some threats to validity of the validation study that needs to be discussed.

In our opinion, the main threat regards construct validity. The students were asked to express their opinions about the course effectiveness in comparison to regular courses. The question here is whether students were able to answer these questions reliably and if they correctly understood the meaning of the categories defined by the

Bloom's taxonomy. Also, the question is hypothetical and refers to students' experience with attending regular courses at the University. Unfortunately, eliminating this threat would require conducting two separate courses (a regular one, and the one in its current form), which was not possible. The other option would be to compare the final grades of the students attending the course with the grades of the students attending previous versions of the same course. Unfortunately, it was also not possible because the program of the course was modified in 2015 when the framework was introduced.

There are also some threats related to conclusion validity concerning multiple testing. We have performed 21 binomial tests for RQ1 and 4 additional tests for RQ2. Therefore, the global significance level was higher than the assumed $\alpha = 0.05$. However, the observed p-values were very small, and even if restrictive correction methods like Bonferroni were used, all but one of the hypotheses that were originally rejected could still be rejected.

Finally, there is a threat related to the generalizability of the proposed framework. Although we currently conduct the third edition of the course, the framework has not been validated in other contexts.

## 5 Conclusions

In this paper, we presented a framework for organizing academic SE courses similarly to Scrum projects. The proposed approach allows for combining lectures and laboratory classes with a minimalistic capstone project divided into a series of iterations (sprints).

The framework proposes a "rolling" Definition of Done that encourage students to use their current sprint of the project to practice the content learned during the lectures and laboratory classes taking place in the previous sprints.

The proposed framework was used to implement a Software Engineering course at Poznan University of Technology. To validate the framework we conducted an end-of-term survey asking students about the effectiveness of the course and impact of the framework mechanisms on their motivation.

In the 2 years of our pilot research, we received 146 responses to our survey. The results show that on average 68% of students agreed that the course organized using the framework allowed them to more effectively achieve the learning objectives related to Scrum on other SE-related content. Also, on average 67% percent of students stated the mechanisms of the framework had a positive impact on their motivation to learn.

**Acknowledgements** I would like to thank the tutors of the Software Engineering II course—Andrzej Jaszkiewicz, Michał Maćkowiak and Marcin Szelag and the B.Sc. students for their participation in the course.

# References

1. ACM/IEEE: Curriculum Guidelines for Undergraduate Degree Programs in Information Technology. Association for Computing Machinery (ACM), IEEE Computer Society (2008)
2. ACM/IEEE: Curriculum Guidelines for Undergraduate Degree Programs in Computer Science. Association for Computing Machinery (ACM), IEEE Computer Society (2013)
3. ACM/IEEE: Curriculum Guidelines for Undergraduate Degree Programs in Computer Engineering. Association for Computing Machinery (ACM), IEEE Computer Society (2016)
4. Anderson, L.W., Krathwohl, D.R., Bloom, B.S.: A Taxonomy for Learning, Teaching, and Assessing: A Revision of Bloom's Taxonomy of Educational Objectives. Allyn & Bacon (2001)
5. Azizyan, G., Magarian, M.K., Kajko-Matsson, M.: Survey of agile tool usage and needs. In: Agile Conference (AGILE), 2011, pp. 29–38. IEEE (2011)
6. Baird, A., Riggins, F.J.: Planning and sprinting: use of a hybrid project management methodology within a cis capstone course. J. Inf. Syst. Educ. **23**(3), 243 (2012)
7. Damian, D., Lassenius, C., Paasivaara, M., Borici, A., Schröter, A.: Teaching a globally distributed project course using Scrum practices. In: Collaborative Teaching of Globally Distributed Software Development Workshop (CTGDSD), 2012, pp. 30–34. IEEE (2012)
8. Keller, J.M.: Development and use of the arcs model of instructional design. J. Instruct. Dev. **10**(3), 2–10 (1987)
9. Kopczyńska, S.: Relating reflection workshop results with team goals. Comput. Methods Sci. Technol. **20**(4), 129–138 (2014)
10. Kopczyńska, S., Nawrocki, J., Ochodek, M.: Software development studio—bringing industrial environment to a classroom. In: Proceedings of EduRex 2012, pp. 13–16. IEEE (2012). doi:10.1109/EduRex.2012.6225698
11. Mahnic, V., Rozanc, I.: Students' perceptions of Scrum practices. In: MIPRO, 2012 Proceedings of the 35th International Convention, pp. 1178–1183. IEEE (2012)
12. Michalik, B., Nawrocki, J.R., Ochodek, M.: 3-step knowledge transition: a case study on architecture evaluation. In: ICSE, pp. 741–748. ACM (2008)
13. Nurkkala, T., Brandle, S.: Software studio: teaching professional software engineering. In: Proceedings of the 42nd ACM Technical Symposium on Computer Science Education, pp. 153–158. ACM (2011)
14. Rodriguez, G., Soria, Á., Campo, M.: Virtual scrum: a teaching aid to introduce undergraduate software engineering students to scrum. Comput. Appl. Eng. Educ. **23**(1), 147–156 (2015)
15. Schwaber, K., Sutherland, J.: The Scrum Guide™. The Rules of the Game. Scrum.org, The Definitive Guide to Scrum (2013)
16. von Wangenheim, C.G., Savi, R., Borgatto, A.F.: Deliver!—an educational game for teaching earned value management in computing courses. Inf. Softw. Technol. **54**(3), 286–298 (2012)
17. von Wangenheim, C.G., Savi, R., Borgatto, A.F.: SCRUMIA—an educational game for teaching SCRUM in computing courses. J. Syst. Softw. **86**(10), 2675–2687 (2013)

# Author Index

© Springer International Publishing AG 2018                                                      221
P. Kosiuczenko and L. Madeyski (eds.), *Towards a Synergistic Combination
of Research and Practice in Software Engineering*, Studies in Computational
Intelligence 733, DOI 10.1007/978-3-319-65208-5